MENTAL MODELS
ALIGNING DESIGN STRATEGY WITH [BEHAVIOR]

Rosenfeld Media
Brooklyn, New York

Mental Models: Aligning Design Strategy with Human Behavior
By Indi Young

Rosenfeld Media, LLC
705 Carroll Street, #2L
Brooklyn, New York
11215 USA

On the web: www.rosenfeldmedia.com
Please send errors to: errata@rosenfeldmedia.com

Publisher: Louis Rosenfeld
Editor: Karen Whitehouse
Production Editor: Liz Danzico
Proofreader: Mary Jean Babic
Indexer: Fred Leise
Interior Design: Allison Cecil
Cover Design: The Heads of State

Library of Congress Control Number: 2007930707

Printed and bound in the United States of America

This book is dedicated to my father, Evert Hale Young, Jr.
His "can do" approach laid the perspective for my entire life.

HOW TO USE THIS BOOK

Necessity is the mother of invention. Historically, tool design—from a sharpened stone ax to a folding Japanese pruning saw—has been inspired by need. The designers themselves range from seemingly isolated genius inventors to cubicle farms of engineers with explicit specification documents. Really successful design—that which solves the problem, is easy to use, and is beautiful to behold while functioning—is hit and miss. Innovation is elusive. Lots of money and hope are put towards products that look encouraging at the outset, but end up not quite reaching the mark. Entrepreneurs, investors, designers, engineers, and customers all get burned more often than they succeed.

There is no one method to follow to create perfect products. But there are many ways to increase the odds. One of them is to understand the reason for the tool in the first place. Deeply investigate what people are trying to get done and line up your solutions to match. Are you trying to solve a small part of a larger puzzle that could be simplified if you look at a broader context of the customer's behavior and philosophies? Do you have so many aspects to your service that it's hard to prioritize where to invest more development dollars?

As first a software engineer and then as a specialist in web applications, I had been doing this broad sort of task analysis for many years before I came up with a way to draw a picture of it all—a mental model, which is an affinity diagram of user behaviors surrounding a particular topic. (See "What is an Affinity Diagram?" in Chapter 1; and see Appendix B (🐘 www.rosenfeldmedia.com/books/mental-models/content/appendix_b) for the story of how the technique evolved.) Mental models help me illustrate a profound understanding of the user, align solutions to areas that make a difference, and chart my way through a decade of development. When people saw the first diagrams, they encouraged me to share them in print. Seven years later, I can finally give you this book.

Who Should Read this Book?

Depending on your role, you might be interested in reading certain chapters first. I anticipate that a lot of you will be practitioners actually creating mental models. I also hope that many of you are responsible for product strategy—directors and executives—and are interested in how mental models can help guide your decisions. For those of you who are project managers and team managers within large organizations, I have included information about how to run a mental model project successfully as a part of a network of other research and design projects. And for all of you who need to persuade key people in your business to cultivate a better understanding of the people who use your products, I am listing key chapters to reference.

PRODUCT STRATEGISTS & EXECUTIVES:

Chapter 1: What and Why? The Advantages of a Mental Model

Chapter 12: Alignment and Gap Analysis

TEAM MANAGERS:

Chapter 2: When? Using Mental Models with Your Other Work

Chapter 3: Who? Mental Model Team Participants

EVANGELISTS:

Chapter 1: What and Why? The Advantages of a Mental Model

Chapter 12: Alignment and Gap Analysis

PROJECT MANAGERS:

Appendix A: How Much Time and Money?

Available on the book site at ⚑ www.rosenfeldmedia.com/books/mental-models/content/appendix_a.

PRACTITIONERS:

Chapter 4: Define Task-Based Audience Segments

Chapter 5: Specify Recruiting Details

Chapter 6: Set Scope for the Interviews

Chapter 7: Interview Participants

Chapter 8: Analyze the Transcripts

Chapter 9: Look for Patterns

Chapter 10: Create the Mental Model

Chapter 11: Adjust the Audience Segments

Chapter 12: Alignment and Gap Analysis

Chapter 13: Structure Derivation

I expect you'll use this book as your resource when you create your first mental model. I also expect this book to be your resource when explaining the benefits of mental models to people in your organization, as you convince them that you really *can* turn the ship around and create user-centered solutions.

What's in the Book?

This book answers some of the harder questions I am asked about how to create and use mental models. I begin with a set of chapters that introduce mental models and talk about *why and when* to make a mental model:

Chapter 1: Why? The Advantages of a Mental Model

Chapter 2: When? Using Mental Models with Your Other Work

Chapter 3: Who? Mental Model Team Participants

The second section describes *the method* used to create a mental model:

Chapter 4: Define Task-Based Audience Segments

Chapter 5: Specify Recruiting Details

Chapter 6: Set Scope for the Interviews

Chapter 7: Interview Participants

Chapter 8: Analyze the Transcripts

Chapter 9: Look for Patterns

Chapter 10: Create the Mental Model

Chapter 11: Adjust the Audience Segments

The third section of the book describes how to *apply a mental model* to your work:

CHAPTER 12: Alignment and Gap Analysis

CHAPTER 13: Structure Derivation

The appendices and bibliography are available as links on the book site at

 www.rosenfeldmedia.com/books/mental-models

What comes with the Book?

This book's companion web site (www.rosenfeldmedia.com/books/ mental-models) is chock-full of mental models-related goodies. You'll find:

- A variety of Excel and Word templates (including ones that help you with recruiting and capturing behaviors)
- Scripts for converting Excel and Word templates into XML Visio and Omnigraffle diagrams
- Every diagram in the book (download and insert them in your own presentations!)
- The book's bibliography
- Appendix A: *How Much Time and Money?* (learn what'll it cost you to develop mental models)
- Appendix B: *The Evolution of the Mental Model Technique* (learn how this method came to be)

The book site also includes a blog where an occasionally lively discussion of mental models breaks out; please join in!

You can keep up with book-related announcements, new content additions, and other changes on the site by subscribing to its RSS feed:

 feeds.rosenfeldmedia.com/mental-models

Here's How You Can Use Mental Models

You might make a mental model for a lot of reasons. For example, you can improve project management by studying those who coordinate project teams. Or, you can invent a new commuting service by understanding all the aspects of how people get to their jobs. You can capitalize on the gaps between the solutions you offer and what your customers are trying to accomplish. You can derive the architecture of your design from the resulting diagram. While the words I am about to list may sound dated when you read this book in a few years, I'll go ahead and say them (so you can envision it more easily in the present day). This method can be used for:

- Digital products, such as internet applications
- Physical products with interactive functions, such as a watch
- Location-aware products, such as a phone
- Methodologies, such as project management
- Information delivery, such as a monthly statement
- Services, such as controlling your household's carbon footprint
- Physical spaces for providing services, such as a library
- Browsable databases, such as knowledge bases
- Platform-specific networked applications
- Online media, online stores, etc.

You get the idea. Throughout this book I use the word "design" to mean something closer to "engineering design"—making something for someone to use. There are tons of other definitions for "design," but for this book I focus on just this one aspect. So when you read the word "design," think of digital, physical, and environmental interactions that people carry out to accomplish something.

Mental model diagrams have been used for design by for-profit and non-profit organizations, universities, government agencies, private individuals, and internal departments. I will illustrate the breadth of applications throughout this book with analogies and real-life examples.

One thing to take from this book is a sense of moving beyond constraints. You're probably not a strict rule-follower, yourself. Just because my background is software design doesn't mean you can't use mental models to develop a government building or a production workflow or anything else you need. Merge the technique with its established cousins in your particular field of expertise, and tell the rest of us how you did it. Treat it kind of like open source: It is yours to manipulate and extend. Let everyone else benefit from your contributions.

My hope is that our generation of designers can execute an inflection point that will be remembered as the point in time when we stopped designing by necessity.

FREQUENTLY ASKED QUESTIONS

What is a mental model?

The top part of the model is a visual depiction of the behavior of a particular audience, faithfully representing root motivations. The bottom part of the model shows various ways of supporting matching behaviors. Where support and behavior are aligned, you have a solution. Where a behavior is not supported, you have an opportunity to explore further. See page 2 for more information.

What if I don't have a big budget?

If your organization already conducts usability tests with some regularity, piggyback short interviews on top of each session. Ask the participant to stay with you for an hour, and spend half the time on the usability test and half on conducting a non-leading interview. See page 35.

What do you mean by "task?"

The word "task" is used loosely. When I use the word "task," it means actions, thoughts, feelings, and motivations—everything that comes up when a person accomplishes something, sets something in motion, or achieves a certain state. See page 133.

What are task-based audience segments?

Task-based audience segments are, quite simply, groups of people who do similar things. While personality types do touch upon behavior, generative research for building mental models requires that you select from groups of people who want to get different things done. Because you will want to tailor your end solutions to fit each audience exactly, grouping audiences by differences in behavior is important. You want to end up with solutions that match behaviors and philosophies closely rather than with one solution that fits several audiences loosely. Figure out what people want to accomplish, look for differences, and group accordingly. See page 46.

How do I uncover the root task?

During analysis, you are required to interpret a little. This is the "art" to the process. You will find it easier if you ask yourself, "What is this person really trying to do?" The idea is to simplify to the "root" task. See pages 135 and 140.

What do you mean by a content map's "content"?

Let me assure you that the name "content" does not limit your map to text documents. Your content map should include all the ways you serve people, including things like monthly account statements or yearly awards banquets, registration for training courses, or a mortgage calculator. See page 234.

Does a content map show every detail of my solution?

It includes all functionality that exists or is intended for your solution. See page 234.

How can analyzing gaps in a mental model show me innovative ideas?

The first thing to look at is the obvious gaps where there is an absence of content items. Your hope is that you can find a gap that you can fill easily. Then look for scarcity of content items. Think about where you can flesh out things a bit. Look for opportunities to redefine, combine, or augment existing content. See page 248.

How can mental models help me make sense of all my web properties?

Each one of your web properties is a building on your internet campus. Each property has its own unique navigation that represents the mental model of the people populating it. See page 214.

TABLE OF CONTENTS

How to Use this Book   iv
Who Should Read this Book?   v
What's in the Book?   vi
What comes with the Book?   vii
Here's How You Can Use Mental Models   viii
Frequently Asked Questions   x
Foreword   xvi

What, Why, When, and Who?

CHAPTER 1
What and Why?
The Advantages of a Mental Model   1
What is a Mental Model?   2
Why Use Mental Models?   9
Confidence in Your Design   9
Clarity in Direction   16
Continuity of Strategy   23

CHAPTER 2
When? Using Mental Models with
Your Other Work   25
Determine Your Research Method   26
How Mental Models Hook into Other UX Techniques   29
Shortcuts and Other Ways to Use Mental Models   33

CHAPTER 3
Who? Mental Model
Team Participants   39
Project Leader   40
Project Practitioners   41
Project Guides   42
Project Support   43

The Method

CHAPTER 4
Define Task-Based
Audience Segments   45
Task-Based Audience Segments   46
Set Research Scope   65

CHAPTER 5
Specify Recruiting Details   71
Estimate the Tally   72
Write the Screener   76
Coordinate Schedules   83
Recruit Participants   87

CHAPTER 6
Set Scope for the Interviews   93
Set Research Goals   94
List Interview Prompts   96

CHAPTER 7
Interview Participants — 105
Chat by Telephone or Face-to-Face — 106
Do Not Lead — 107
Plan Your International Interviews — 125

CHAPTER 8
Analyze the Transcripts — 131
Comb for Tasks — 132
Get Some Practice — 154
Answers — 162

CHAPTER 9
Look for Patterns — 163
Group Tasks into Patterns — 164
Plan Your Logistics — 189
Congratulate Yourself — 195

CHAPTER 10
Create the Mental Model — 197
Build the Model Automatically — 198
Build the Model Block-by-Block — 200
Review the Diagram with Project Guides — 216
What Did You Learn? — 218
Decorate the Diagram — 218
Ask for Feedback — 223

CHAPTER 11

Adjust the Audience Segments   225

Compare Results to Original Hypothesis   226
Clarify Segment Names   227
Adjust Segment Definitions   228
Use Audience Segments for Other Projects   232
Transition from Research to Design, Verbs to Nouns   232

Applications

CHAPTER 12

Alignment and Gap Analysis   233

Draw a Content Map of Your Proposed Solution   234
Align the Content Under the Mental Model   237
Consider the Opportunities   248
Share the Findings   254
Print the Diagram   257
Prioritize the Opportunities   258

CHAPTER 13

Structure Derivation   265

Derive High-Level Architecture   266
Provide Vocabulary for Labels   276
Test Your Structure and Labels   280
Generate Features and Functionality   280

Index   285
Acknowledgments   296
About the Author   299

FOREWORD

"You're researching all the creativity out of this project!"

I can't tell you how many times I've heard designers, developers, and even business owners say this. It usually comes just after a project has begun, as I'm preparing for interviews with users. Designers just want to start designing, developers want to start writing code, managers want the thing to ship—so why are we spending all this time talking? And all this stuff just seems so obvious. Do we really need users to tell us what we already know?

I try to be diplomatic. "Maybe a few interviews now will save us lots of grief later," I tell them. "Think of this as insurance: Let's make sure we've got the basics right before we've designed everything and written all the code."

But no matter what I say to convince a team to do research early in their project, I never let them know my dirty little secret: I used to be just as skeptical as them.

I've always believed in a user-centered design methodology. Even early in my career, when I was journalist, we always started with the mantra "know your audience." Later in my career, I'd go to conferences and watch presentations with process diagrams—boxes representing users needs with arrows pointing to boxes representing product requirements. Intellectually, I agreed. But when I started a new project, in that intoxicating first stage when anything is possible, I'd jump straight to solutions. "Let's use Flash for this part! And over here, we'll design some awesome icons for navigation..." Our users were still important, but they were there to bear witness to how cool our designs were.

Then I met Indi Young. Indi and I were among the founding partners of Adaptive Path, a user experience consulting company that focused on research-driven design. We founded the company in the dark days of the Web industry. It was 2001. "Dot com" was a dirty word, companies were cutting their Web budgets, and projects were drying up everywhere.

It was then that "research-driven" started having real meaning to me. As Indi introduced her methodology and resulting visualizations, it became clear that she wasn't just trying to make designs better in some abstract way. Rather, her process was simple enough to resonate with anyone on a Web team. And perhaps more importantly, it would help connect Web teams to other core parts of their organizations who were skeptical of spending even another cent on their web sites.

In the end, using Indi's process, we were able to convince teams that we weren't researching all the creativity out of their projects. We were researching the risk out. And no matter how the industry is faring, that's a story people want to hear.

This book is an excellent guide to a research method firmly grounded in common sense. But don't let the simplicity of the process detract from the power of the change it can enable. Talking to users in a structured way, analyzing in a collaborative way, and diagramming with clarity can transform the way you approach the Web.

And it might just ignite your creativity!

Jeffrey Veen
San Francisco
August, 2007

CHAPTER 1

What and Why? The Advantages of a Mental Model

What is a Mental Model?   2
Why Use Mental Models?   9
Confidence in Your Design   9
Clarity in Direction   16
Continuity of Strategy   23

1

You might be thinking, "What does she mean by 'mental models?'" Since the phrase "mental model" is somewhat commonly used—at least in the realm of research—I want to set out what I mean by the term and then outline why you would ever want to make one.

What is a Mental Model?

"The deepest form of understanding another person is empathy...[which] involves a shift from...observing how you seem on the outside, to... imagining what it feels like to be you on the inside."[1]

Designing something requires that you completely understand what a person wants to get done. Empathy with a person is distinct from studying how a person *uses* something. Empathy extends to knowing what the person wants to accomplish regardless of whether she has or is aware of the thing you are designing. You need to know the person's goals and what procedure and philosophy she follows to accomplish them. *Mental models give you a deep understanding of people's motivations and thought-processes, along with the emotional and philosophical landscape in which they are operating.*

Mental models embrace anything from looking up a part number online to asking the guy at the hardware store how to mix epoxy. A mental model consists of several sections, with groups within each section. Mental models are simply *affinity diagrams* of behaviors made from ethnographic data gathered from audience representatives.

---

[1] From the book *Difficult Conversations* by Douglas Stone, Bruce Patton, and Sheila Heen of the Harvard Negotiation Project, Chapter 9, "Empathy is a Journey, Not a Destination," p. 183.

What is an Affinity Diagram?

Affinity diagrams, in the simplest interpretation, show groups of related things. You can make an affinity diagram out of your grocery list. Some items, like milk and eggs, might be found near each other in your store. Other items, like pasta sauce and spaghetti, are elements of a single meal you're planning. The diagram below shows an example.

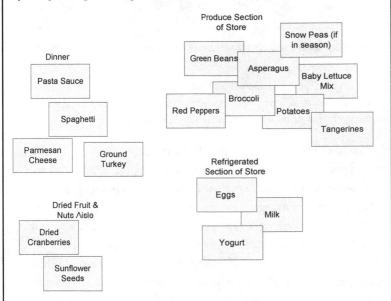

A mental model for a particular topic is, in essence, an affinity diagram of user behaviors. The towers in the diagrams represent group names for the behaviors within. The sets of towers represent a higher-level group of the behaviors.

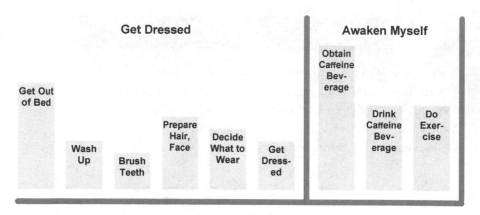

FIGURE 1.1.
Mental model of a typical morning for people who commute to work or school. There are additional examples on the book site under Cases: 🚌 www.rosenfeldmedia.com/books/mental-models

For example, when you wake up in the morning you get dressed, you eat, and you get on the train. These can be considered "mental spaces" in a diagram of your morning (Figure 1.1). On holidays you skip the "get on a train" mental space and instead you "eat a big breakfast with the family." On mornings when you are tired, maybe you add a mental space about "become awakened" by perhaps drinking coffee or tea or doing some exercise. So the full mental model about your morning has several parts. The "Eat" section would have various divisions within it depending on whether you were heading to work or joining the family for Sunday brunch.

To create a mental model, you talk to people about what they're doing, look for patterns, and organize those patterns from the bottom up into a model. From the field research, you will glean maybe 60 or 120 behaviors per person. Over time you see the same behaviors and you group them together. You line them up in towers; then line up the towers into groups that represent different cognitive spaces. The diagram looks a lot like a city skyline.

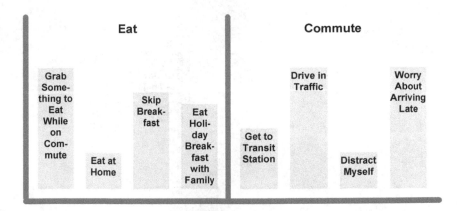

Once you have created the top half of the diagram, you focus on the lower half. Take the product features that you intend to create and align them beneath all the towers they support. In other words, you align the features that your business values beneath concepts that people mentioned. When you have finished, you will see areas of the mental model that are less supported than others, and you may have leftover functions that don't support anything in the mental model. The resulting diagram tells a story about the viability of your business strategy for a particular solution. In Figure 1.2, dark green indicates a primary match for the feature. Light green indicates additional secondary matches for the feature. In other words, for every light green feature there is one dark green feature aligned beneath the best match. Excess features that do not map to the mental model appear in the lower right corner.

Use the name "mental model" whether the diagram shows just the towers above the horizontal line or it shows the features aligned beneath the towers. It is this entire picture that becomes the heart of your strategy.

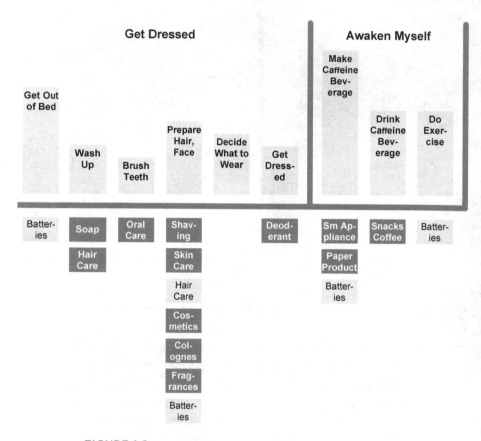

FIGURE 1.2.
Mental model with features aligned beneath it.
(Features borrowed from the product category list from
Procter & Gamble's site www.pg.com)

Taking the top and bottom half together, the resulting mental model is a
diagram of how a certain segment of people tend to accomplish something,
with the things you are making aligned to the depicted concepts. *You use
the model to understand how your current offerings do and do not support
people and devise your strategy going forward.* You do

this through multiple workshops with team members and stakeholders in your organization, which develops understanding and innovation. The model has a long lifespan, so you can use it to direct your progress with deep awareness of user-centered design for 10 or more years.

The Mental Model Process

First, reach out to actual users and have a conversation with them, collecting their perspective and vocabulary. Analyze all of those conversations and composite them into a diagram called "the mental model diagram." Then compare all of the things your solution is supposed to do with the different parts of that mental model diagram. Align them with the concepts that they support. You can do this with functionality just as it exists, or functionality being planned, or you can play around with brainstorming new ideas. When you step back and look at the whole picture with teammates and stakeholders in the organization, you can develop a design strategy—a vision—to follow over the next decade. Then you can start devising tactical solutions for high priority areas of the mental model.

The mental models defined in this book are models of a person's somewhat stable behaviors, rather than ephemeral models that are temporary representations of one situation. I want to acknowledge this distinction because those in the field of cognitive research have explored mental representation in great detail in the past decade, and I want to indicate where these mental models might fall within the currently defined parameters. "'Mental model' has become a more generic term for mental representation. Cognitive research is now so specialized that article abstracts begin with verbose strings of qualifiers to narrow down the type of mental representation they mean."[2] Because the mental models in this book are collections of the root reasons why a person is doing something, they belong to the set of mental representations that are built over a long period of experience and are thus resilient. These mental models represent what a person is trying to accomplish in a larger context, no matter which tools are used.

---

[2] Jay Morgan, Applied Cognitive Scientist, MS in Applied Cognition and Neuroscience, University of Texas at Dallas, 2004.

Why Use Mental Models?

"Why should I use a mental model?" This is probably one of the questions that prompted you to open this book—indeed, it's a good one.

Using a mental model can advance several tasks for you—both from a tactical and a strategic standpoint. It can guide the design of the solution you are working on. It can help you, and your team, make good user and business decisions. And, it can act as a roadmap, ensuring continuity of vision and opportunity as the makeup of your team evolves over the next decade.

The Three C's

You might notice that the three main reasons I use to describe the advantages of mental models all begin with the letter "C":

- *Confidence* in Your Design—guide the design of the solution

- *Clarity* in Direction—make good user and business decisions

- *Continuity* of Strategy—ensure longevity of vision and opportunity

I thought this was a neat way to remember the reasons, especially if you have to persuade your CEO in the elevator why you want to create some mental models of your customers.

Confidence in Your Design

How do you know if you've got it right? You're looking for something that will ensure that you've hit the mark. A mental model will give your team members confidence in their design because it is based on a solid foundation of research. It will assure management that success is likely. Likewise, your users will have confidence in using the design because it matches what they already have in mind. They will not hesitate while using your solution. It will make sense to them, embody some of their philosophies, and respect the emotional component of what they are doing.

Leverage Luck + Intent

As you might suspect, any number of variables combine to make success— not the least of which is luck. "Being in the right place at the right time," experiencing "a freakish alignment of the stars," "having things just go right"...no matter the phrase, luck plays a larger part in the process than any organization wants to face. Michael Bierut, a respected visual designer, said this about luck during an interview[3] with Adaptive Path founder Peter Merholz: "It's a dirty secret that much of what we admire in the design world is a byproduct not of 'strategy' but of common sense, taste, and luck. Some clients are too unnerved by ambiguity to accept this and create gargantuan superstructures of bullshit to provide a sense of security."

You're fortunate if you work in an environment where the "luck component" is acknowledged. Most people have to justify their decisions with cold, hard facts. What I like to bring to the picture is a tool that can free you to recognize possibilities[4] while at the same time provide solid data. In other words, you can "embrace the ambiguity"[5] of the design process because you have a mental model to steer you.

---

3   "A Conversation with Michael Bierut," tinyurl.com/pkruo

4   Adaptive Path blog entry "Designing for Luck," tinyurl.com/2hh9oc

5   Favorite mantra of Mary Piontkowski, researcher, information architect, interaction designer, and a frequent collaborator of mine, and contributor to mental-model process adaptations and improvements.

Scientific versus Intuitive Methods

The mental model method is a qualitative approach based on interpretation of data that looks like a scientific method. It is a hybrid produced by science and intuition; it's a little of both. It is a very successful method in environments where people are looking to support decisions with real data. It is also enormously useful in environments where teams can define and communicate product/information design with more intuitive techniques such as storyboards, comics, or videos.[*]

---

[*] See the work of Kevin Cheng and Tom Wailes at Yahoo!, as presented in "Finding Innovation in the Five Hundred Pound Gorilla" at IA Summit 2007. tinyurl. com/38wdbn. Also see Jared Spool's June 2007 UIE article "Knowledge Navigator Deconstructed: Building an Envisionment" tinyurl.com/ywsx7m

So how does a mental model give you the evidence you need to support your design, in addition to the leeway to create luck? A mental model is a visual language. Its text is the data. Its grammar is the vertical and horizontal alignments of concepts. With a language you can convey anything you can think of. Frank Gehry, the well-known architect of the Walt Disney Concert Hall in Los Angeles, explains[6] it this way, "For Disney Hall, I spent a lot of time thinking about how to listen. I worked closely with the acoustician, who said sometimes the sound has to be big, and then sometimes it has to be like candlelight. I got a sense of what a conductor thinks about when he gets up on stage, and what the musicians need in their relationship to the room. Once I had all of that, I could free-associate, because I was programmed like a computer. I couldn't go wrong because I'd learned the language so completely. It's what I try to explain to my students—the more you know, the freer you are in the end."

---

[6] *House Beautiful*, June 2006, "Giants of Design" awards article, page 120.

People have had similar experiences with the language of a mental model. Jeff Veen, User Experience Manager at Google and founding partner of Adaptive Path, created a mental model diagram for his redesign of Google Analytics in 2006. He says he used the mental model as a way to gain empathy with the people who were trying to understand their web traffic. Talking to users "bought me time to absorb the difficulty of the problem we were facing and time to start pondering solutions to that [problem]." Although he did not use the diagram to specifically derive the design, he says, "Most of what made it into the design was based on my own ...empathy [which] was certainly developed while going through this research and analysis."

Distinguish Among Solutions

A mental model represents the entirety of each audience segment's environment. Thus, the diagram depicts where one segment's experience ends and where the next one begins. You certainly wouldn't want to combine the experience of truck drivers with that of dispatchers—that much is obvious, and the separate mental models will show it. More subtle is the experience differences among people of a similar segment but, say, in a different country. Do the differences merit separate solutions or not? Compare the mental models. If they have a lot in common, a common solution is dictated. If not, then different solutions are needed, each with its own architecture matching the mental model. As another example, corporate intranets should be specific to each workgroup type. Software engineers need a different set of tools on their intranet than corporate lawyers. Understanding the differences among the mental spaces of your audience segments will bring clarity to your design.

Assemble Original Ideas

Use the mental model diagrams to derive design decisions. For example, if a tower in the diagram shows that people "Collect Pictures of Renovated Kitchens to Mull Over," perhaps you can create a scrapbook of renovated kitchens for a showroom, or point people to "kitchen renovation" as a tag word on Del.icio.us. Use the person's real world as inspiration. What if there is a tower called "Distrust Sales Reps"? What can you possibly do to help customers who believe that talking to a sales representative is a waste of time? Connect them directly with the technical representative? Get rid of sales representatives entirely, and change the way your organization approaches potential customers? Deriving ideas from the diagram will simplify the work of deciding which features support which behaviors. Prioritizing these feature ideas according to business goals and resources will simplify discussions about what the customer intends to do and how the business will serve them.

Validate That Ideas Match Needs

You can use the mental model to double-check design decisions, just as you use personas to do a mental double-check: "What would Meredith do?" Say someone powerful in your organization decides he really wants a stock ticker to appear on a particular web site. You can validate that request by matching it to a tower in the mental model that shows "track market stock prices." If no behavior of this description appears in the mental model, you can respectfully point out this absence and possibly talk the person out of an unnecessary feature.

How Did Mental Models Help Your Organization?

"The mental model provided us a global, durable definition of tasks, processes, terminology, and guiding principles to design our intranet HR portal and quickly course-correct whenever teams propose new programs with obscure names, hidden in odd parts of the architecture, whose value would not be obvious to our employees."

—Jacqueline De Muro, e-HR Channel Manager, Global HR Services, Agilent Technologies, Inc.

Avoid Politics

Who among us has not worked in a situation where other people wanted to second-guess your design decisions? Sometimes these people can comprise whole departments. Battles ensue among groups as to who makes the final design decision, or who gets preferential placement in a particular solution (e.g., placement in primary navigation of a web site). That's office politics. A mental model can act as a third-party mediator. It is essentially a collection of data placed in relationship to other data based on your interpretation. The data is concrete—your interpretation is circumstantial. Suddenly the conversation isn't about the design you've created; it's about your analysis of the data. Conversations with your erstwhile foes center on phrases like, "My understanding of what this customer means is x; what is your interpretation?" People from different disciplines can come to a common understanding of the customer's situation. You both end up on the same side of the table, both looking at the neutral data. This fortuitous arrangement produces more effective design discussions and faster decision-making.

How Did Mental Models Help Your Organization?

"Developing a design using mental models (and displaying the results) is a durable reminder to the team and the organisation that you did not just 'make something up.' It's the IA equivalent of calculus from first principles. Once you have done it, it's hard to imagine justifying design any other way."
— Craig Duncan, Head of the Information Management Unit of the U.N. International Strategy for Disaster Reduction

Derive Architecture

If you are designing for software, you can use the complete diagram to derive the top-level organization of the application, such as navigation or toolbars. Additionally, you can use the mental model as a starting point for interaction design and as a guide for feature definition.

How Did Mental Models Help Your Organization?

"The framework provided by a user mental model takes the guesswork out of IA, or at least makes your guesses more likely to be on target."

—Camille Sobalvarro, Senior Manager of Web Marketing, Sybase

Stop Spending All Your Time Re-Architecting

I see a lot of companies who are still stuck in a cycle of redesign—re-architecting what they have created over and over. They define something to fix, change it, and then realize the initial definition was a little off. I've seen many five-year roadmaps that look like the one in Figure 1.3, but all these important items keep getting pushed off because the architecture still isn't right or the strategy keeps changing.

Five-Year Plan

- Improve functionality of tools

- Add better tools to empower the customer

- Serve dynamic news, events, announcements based on region and role

- Fine-tune messaging (marketing) opportunities (internal ads, external ads, contests and giveaways, special sales, landing pages for brand-specific islands)

FIGURE 1.3.
Five-year roadmap for one web property belonging to a multi-national semiconductor corporation.

Don't stay stuck in this cycle of re-inventing your solution or re-addressing your information architecture. You need an overhaul of your regular development process. Ryan Freitas of Adaptive Path refers to this as "punctuation in product development."[7] Use a mental model as a guide to

---

7 Future of Web Design conference, London, UK, April 2007, www.futureofwebdesign.com/

get the definition right the first time, then focus your energy on fulfilling business goals instead of focusing on iterative changes within a single medium.

Clarity in Direction

The clarity argument brings your internal organizational structure and its strategies into the picture. You need a clear design strategy that will act as a container around all of the solutions you create.

Pay Attention to the "Whole Experience"

The "whole experience" includes all the ways an organization interacts with its users: stores, account statements, customer service calls, product ordering web sites, packaging, and so forth. Jared Spool, the founder of User Interface Engineering, writes a comparison[8] about whole experiences between MP3 players from Apple and SanDisk. He says, "SanDisk hasn't created what Apple has built: a powerful user experience for listening to great music. While possibly technologically deficient, the iPod combines the player hardware with the iTunes software, the iTunes Music Store service, the Apple stores for sales and support, and the prestige that comes from the Apple brand...SanDisk can't compete if it only focuses on the hardware engineering." Businesses that pay attention to the entire spectrum of customer interaction, and get it right most of the time, win attention and loyalty. Because the mental model depicts the whole of the user's environment—it is not focused on one aspect, service, or tool—it represents the user's perspective of the whole experience.

---

[8] Jared Spool's UI11 conference article "Innovation is the New Black," tinyurl.com/olwgh

Emotion in the Experience

There is a lot of lingo with X in it these days: UX (user experience), MX (managing experience), and UX strategy. The X stands for "experience," which has to do with the whole environment in which a person interacts with your solutions and your organization. The concept of "the user experience" has been around for a long time, especially in disciplines other than human-computer interaction, such as architecture and retail store design.

For an experience to be considered successful, people have to *be able* to use it and *want* to use it. I've heard this described as useful, usable, and desirable. It applies to all sorts of fields—toys, buildings, silverware, electronic devices, etc. Here is a quote from a Swedish furniture showroom owner that sums this up nicely. "Don't make something unless it is both necessary and useful. But if it is both necessary and useful, don't hesitate to make it beautiful."[*] There is an emotional component to the use of a thing that businesses are becoming more aware of.[**] Mental models capture not only the cognitive intent of a person but also the emotion, social environment, and cultural traits of a concept. The alignment of possible business strategies completes the picture.

---

[*] *House and Garden* magazine March 2007 article, "The Design-Obsessed Traveler—Stockholm," quoted as "the credo of Design House Stockholm, a top city showroom."

[**] See Donald Norman's book *Emotional Design: Why We Love (or Hate) Everyday Things.*

Peter Merholz, one of the founders of Adaptive Path, puts it this way. "An experience strategy is a clearly articulated touchstone that influences all the decisions made about technology, features, and interfaces. Whether in the initial design process, or as the product is being developed, such a strategy guides the team and ensures that the customer's perspective is maintained throughout."[9] He also strongly believes that you should stop designing

---

[9] *BusinessWeek* online article "Experience Is the Product" by Peter Merholz on June 22, 2007. tinyurl.com/2cqcbz

products. "When you start with the idea of making a thing, you're artificially limiting what you can deliver...Products are realized only as necessary artifacts to address customer needs. What Flickr, Kodak, Apple, and Target all realize is that the experience is the product we deliver, and the only thing that our customers care about."[10]

Use Design as a Business Advantage

Brandon Schauer, a design strategist at Adaptive Path, has been talking[11] about the increasingly important role of design in businesses. He says because competitive advantage is shrinking, the focus has started to shift towards what kind of top-line value can be gained for the business, both internally and externally. He references a seminal 1996 essay written by the celebrated Harvard Business School professor of competitive strategy, Michael Porter, which stated, "A company can outperform rivals only if it can establish a difference that it can preserve. It must deliver greater value to customers or create comparable value at a lower cost, or do both."[12] Brandon is exploring what levers exist and can be adjusted. In 2005 Brandon interviewed innovation and strategy consultant Larry Keeley of Doblin, Inc., who had this to add: "A growing number of business [and government] leaders...have come off of two decades of trying to find greater efficiencies...And the evidence is overwhelming that that has worked. But I think it's equally clear to good leaders that they can't continue to expect...massive improvements in efficiency each and every year...Now most good leaders are saying, 'I've got to figure out a way to get to organic growth; I've got to figure out a way to do something powerful and new.'...[A]nd unless they're newsworthy, and unless they're

---

[10] "Experience IS the Product...and the only thing users care about" by Peter Merholz in June 2007 for the Industrial Design Supersite Core77, tinyurl.com/2dtrvb

[11] Brandon Schauer presented in "Connecting Design to Real Business Value" at Adaptive Path's Managing Experience through Creative Leadership conference, in San Francisco, February 2007. Slides at tinyurl.com/3bg3c7. Brandon has a Master of Design degree from the Institute of Design in Chicago and an MBA from the Stuart School of Business. For more of Brandon's writing, see his blog at www.brandonschauer.com/blog

[12] Harvard Business Review, "What is Strategy?" by Michael E. Porter, November 1996. tinyurl.com/ypqzfs

startling, and unless they really compel customers, they tend to fail."[13]
So Brandon's point is about marrying this focus on top-line value with
attention to the whole experience a customer has. Do this and you have a
new competitive advantage.

Experience Strategy

The strategy that you develop for your product ought not evolve in isolation.
Even though the value of user experience* is clear, your over-arching rea-
sons for providing something should be considered with equal weight. Jesse
James Garrett describes the phrase Experience Strategy thusly:

$$\text{Experience Strategy} = \text{Business Strategy} + \text{UX Strategy}^{**}$$

A mental model helps you visualize how your business strategy looks com-
pared to the existing user experience. Thus, it is a diagram that can support
your experience strategy.

---

* See the diagram by Bryce Glass, "The Importance of User Experience," from March
2006: tinyurl.com/ysmcrn

** Jesse James Garrett in his introductory address to MXSF 2007, "Experience Strategies
— The Key to Long-term Design Value." See Jesse's blog at blog.jjg.net

---

Evolve Your Organization

Perhaps you need to transform your organization into an entity that pays
more attention to user experience. You need a strong tool to assert change
within your organization or get the attention of people who can spread that
change. The sheer visual force of a mental model, covering several feet of
a wall when pinned up, is enough to make people stop and look. They are
easy diagrams to read and understand. Post them everywhere you think
people might have time to glance through them. Post them outside your

---

[13] Institute of Design, Strategy Conference Perspectives 2005, Issue 2, "Interview: The
Emergence of New Innovation Disciplines," Brandon Schauer interviews Larry Keeley.
tinyurl.com/2ajpeb

work area, in popular conference rooms, in the lunch room, even outside the bathrooms. Jacqueline De Muro's team at Agilent Technologies, Inc. occupied a building with limited wall space, hence her inspiration for placing the diagrams outside the bathrooms, which was the only space available where people would come and go on a daily basis. I encourage you to post them in other departments and in other buildings, where someone you've forged a relationship with can sanction their presence. Any opportunity to break down the wall between your organization and your customers is precious. Similarly, the opportunity to reach across chasms among departments is priceless.

If your team uses the sticky-note approach during analysis of the data, you will have large surfaces covered with little square paper notes. Craig Duncan of the United Nations in Geneva, Switzerland, has created a few mental models. He says, "The walls of sticky notes create a perfect opportunity to get management and other colleagues to engage (and be impressed by the process). I strongly recommend holding a TGIF or other small office party while the Post-Its™ are all up. They start conversations, and convert people to the fact that IA is indeed a legitimate discipline."

How Did Mental Models Help Your Organization?

"It helped us talk *to* our users rather than about them."

—Simonetta Consorti, Information Architect on the UN Prevention Web Team

Adopt the Customer's Perspective

In the past couple of decades a paradigm shift took place in large corporations towards the philosophy of putting the "customer first." This paradigm traditionally applies to service and support, but also has been applied to product design. Companies are finding that as global competition increases, fewer people are beating a path to their doors. Rather than creating a product internally, and then hiring marketing and sales people to persuade customers to buy it, companies are taking a closer look at what customers need. They are designing products that will sell on their own merit.[14]

How Did Mental Models Help Your Organization?

"Mental models are like user perspective goggles."

—David Poteet, President of New City Media

I often work with groups who have adopted the "customer-first" philosophy but don't quite have the vocabulary for it. Employees in these organizations tend to answer questions about customers in terms of internal business goals. At the beginning of each project, I interview all the people who have a stake in the project's outcome. I ask each of them: "What benefits will your customer see from this project?" Most stakeholders will respond with an answer that morphs from a vague customer problem directly to their business goal. To protect the innocent, I'm not going to name names. Here is an example paraphrased from what I hear: "Right now, the navigation on the site is really just a bad experience. We'll make the experience a lot easier for them. Rather than generate lots of content, we're going to be smarter about it and help them filter it. Customer satisfaction scores will go through the roof."

This echoes what they're thinking from an employee focus; therefore, the answer is from a business perspective. What I hope to achieve by developing task-based audience segments and mental models is to

---

[14] Jeff Veen of Google talks about designing from the user's perspective during an interview with Josh Owens and Chris Saylor on the *Web 2.0 Show*, tinyurl.com/ywaonn. For more of Jeff's writing, see www.veen.com/jeff/index.html.

persuade business stakeholders to speak at length from the customer perspective. I want to hear these employees answering the question with something more like this: "As a customer, I will get appropriate content that is relevant to my problem. I won't have to sort through so many content choices. I will be able to quickly filter all the information for what I need."

Adopt the Customer's Verbs

To start this subtle paradigm shift, I encourage everyone in the organization to start speaking in verbs. People are usually already aware of the customer perspective and ask questions such as, "Well, what is the customer trying to do?" The answers often echo the question: "The customer is trying to deploy the system." There is peril in this approach. The "customer is trying to" phrase puts a barrier in place. It telegraphs that the speaker is someone other than the customer. It means that the speaker is describing his understanding of what that customer is doing. I ask people to cut to the essence. Most companies know something about what their customers are doing; state these actions confidently. Use verbs that describe actions from the customer's point of view. Choose expressive verbs that are representative of a specific situation. If the customer really is "trying to deploy the system" because they are not sure they can actually deploy it, the illustrative verb "attempt" might be a better choice. These verbs naturally have to come from the customer's point of view rather than the business employee's. Either "Figure out how to deploy the system" or "Attempt to deploy the system" is a better choice. Or perhaps there is no problem deploying the system, so "Install the system" or "Roll out the system" may better describe what the customer is doing.

Being inside your customer's head is powerful. You can see how using the customer's verbs expresses their view of the world. Try to be aware of the way you speak to colleagues at meetings. Use the customer's verbs. Verbs are the most powerful way of getting people to shift towards "customer first" for product design.

Verbs from the Customer

Katie is a business user of a suite of office software created by a certain well-known Seattle manufacturer. Here are some of the verbs she uses to describe her work day:

- Learn details of using company products
- Write help guides for company products
- Read cases from customer support
- Argue for product improvements to software team
- Recommend marketing approaches to marketing team
- Study XML manuals for writing XML-based help guides
- Attend XML training
- Create help guide formats/templates in Word
- Organize Word files
- Submit bug to software team
- Help co-worker format his Word templates
- Look up Word formatting question

These descriptive verbs help you see Katie's world from her eyes.

Continuity of Strategy

A mental model with features and solutions aligned beneath it becomes a roadmap for strategy over the next decade. After your initial brainstorm of how to support users, you will know that you can't possibly implement every idea in the next three or four quarters. Things will get pushed off. New ideas will materialize as the market and technology changes. You will see new opportunities in three years that don't exist today. The diagram persists as a visual map of where you plan to go.

Rely on Mental Models to Change Slowly

User mental models change slowly. Take the concept of cash. Coinage has been around for centuries upon centuries. People exchange coinage for services, goods, and materials, and they have developed ways to carry, obtain, and securely store cash. Then the ATM came along and changed how people obtain cash. It has not yet been fully adopted. Many still prefer the mental model of going to a bank and getting cash from a teller. Now plastic exchange of currency has become widely available. You can use a debit, credit, or gift card in place of currency, which makes the act of carrying, obtaining, and securely storing cash largely unnecessary. But adoption of the plastic model was slow, thus the mental model regarding cash was slow to change. You can be reasonably sure that most mental models you make will likewise be valid for many years to come. Basing a continuous strategy on a long-lived artifact is a good idea.

Keep the Knowledge, Shift the Team Members

Also, it is a fact of life that the members of a particular team change over time. One of the more difficult problems organizations face is preservation of internal knowledge. Mental models guide your team's progress over the years and become a place where decision history and rationale is recorded, as a foundation of decisions to come.

So if you're thinking a mental model might help you with your work, you may be asking, "How does it fit into what I'm already doing?" The next chapter will answer this question.

CHAPTER 2

When? Using Mental Models with Your Other Work

Determine Your Research Method   26
How Mental Models Hook into Other UX Techniques   29
Shortcuts and Other Ways to Use Mental Models   33

Mental models are just one tool in your toolbox. Here's an explanation of how they work in conjunction with other methods.

Determine Your Research Method

The best way to choose a user research method is to know what each technique is good for. Most user research techniques can be categorized into three groups: preference, evaluative, and generative. The first of these groups is the most widespread—preference. Not only do you see marketing departments sending out surveys and conducting focus groups, but you also see opinion polls flourishing in the media. Preference research is the most common type of customer research. It is perfect for canvassing a large number of people to determine how the product will be accepted or preferred by people. In Figure 2.1, you will see a sampling of techniques that support preference research. In the last column, there is a set of uses for preference research, such as branding. You'll notice that you don't use preference research to design interaction or information architecture. And, you don't use it to find out how well a solution works.

In the second row of Figure 2.1 you will see evaluative research. In the past few years, it has become more common to hear decision-makers call for product testing before releasing an item to the market. This request is also fairly frequent when the "market" is a large internal organization and releasing an efficiency tool can make a difference in the bottom line. In the table, you will see that "card sort" and "customer feedback" appear for evaluative research as well as preference research. Card sorting[1] can be used for a wide variety of purposes. The feedback you get from customers also ranges from complaints to opinions about how to fix your product. Note that the uses for evaluative research are specific: You can improve existing interaction functionality and the visual design of the screen layout for software, hard goods, or a service. You also can find out which labels, nomenclature, and instructions make the most sense to the user.

---

[1] See Donna Maurer's upcoming book *Card Sorting*, published by Rosenfeld Media. www.rosenfeldmedia.com/books/cardsorting/

DATA	TECHNIQUE	USES
Preference Opinions, likes, desires	Survey Focus Group Mood Boards Preference Interview Card Sort Customer Feedback	Visual Design Branding Market Analysis Advertising Campaigns
Evaluative What is understood or accomplished with a tool	Usability Test Log Analysis Search Analytics Card Sort Customer Feedback	Interaction Functionality Screen Layout Nomenclature Information Architecture
Generative Mental environment in which things get done	Non-Directed Interview Contextual Inquiry Mental Model Ethnography Diary	Navigation & Flow Interaction Design Alignment & Gap Analysis Contextual Information Contextual Marketing

FIGURE 2.1.
This User Research Types matrix will help you decide which research to use.

Generative research has been around for a long time, though its application to software strategy is more obscure. Generative research explores the mind space of someone doing something. It is research that is conducted before the ideation phase.[2] It focuses on a higher level than evaluative research, asking the end purpose for every tool used rather than the details of how well a specific tool is applied. Open-ended research methods, such as ethnography, non-directed interviews, and diaries, allow researchers to create a framework based on data from participants. This framework then can be used to guide information architecture, interaction design, and contextual placement of information and products.

---

[2] See the work of Liz Sanders at maketools.com

For example, you would use generative research to find out how people buy books, which usually doesn't differ based on age, gender, interests, or preferences. You would use preference research to find out *which* book a person would buy. However, the act of buying is not that different: Look for the book, make sure it's the right choice, and then purchase it. Mental models generated with generative data and aligned with proposed information and functionality can deliver an unambiguous picture of how well a solution supports the user through gap analysis.

Building Products Based on Preference Research is Like Building a Kitchen from a Stack of Magazine Clippings

Imagine you are an architect talking to a couple who wants a kitchen re-model. They bring you a stack of magazine clippings with photos of kitchens they like. They talk about how they want to cook gourmet meals for friends with fresh produce from the farmer's market. Your next step is to start drafting blueprints based on all this input, plus the information you already know, such as the efficiency of a work triangle in a kitchen. But imagine if you were not allowed to draw blueprints, and instead you were required to hand over the stack of magazine clippings to the contractor. Assume the contractor has never built a kitchen before, so he has no experience with kitchen functionality and work triangles. Without your skills at interpreting client input into a remodeling plan, the project would stall.

Asking engineers to build a product based on a stack of preferences is just like asking a contractor to build a kitchen based on magazine clippings.

Stepping back to a more general, rather than user/consumer, definition of research, Wikipedia defines three main forms of general research methods:

- Exploratory research, which structures and identifies new problems
- Constructive research, which develops solutions to a problem
- Empirical research, which tests the feasibility of a solution using empirical evidence[3]

In a way, mental models embody both exploratory and constructive research, allowing you to derive solutions to problems from the data set as well as structuring where new problems for the next year might lie.

How Mental Models Hook into Other UX Techniques

If you are who I think you are, you're familiar with standard user-centered design techniques, such as writing scenarios based on specifically designated personas. You've probably seen or used affinity diagrams that show group relationships among things. You may have commissioned or participated in a field study of your users. You possibly have directed "Voice of the Customer" surveys as a part of a Six Sigma[4] program at your company. But you're looking for something to pull these techniques together, to make them reach further.

Mental model research occupies a place in the constellation of techniques after user data collection and before product and interaction design concepts (see Figure 2.2). Its use as a planning roadmap is long-lived. You can refer to the same mental model for several projects over time.

---

3 Wikipedia: en.wikipedia.org/wiki/Research

4 A brief explanation of Six Sigma appears on my book site under the Resources section:
   www.rosenfeldmedia.com/books/mental-models/content/resources For further exploration of many other user research techniques, there is a fabulous matrix of these tools in June Cohen's book, *The Unusually Useful Web Book*, page 49.

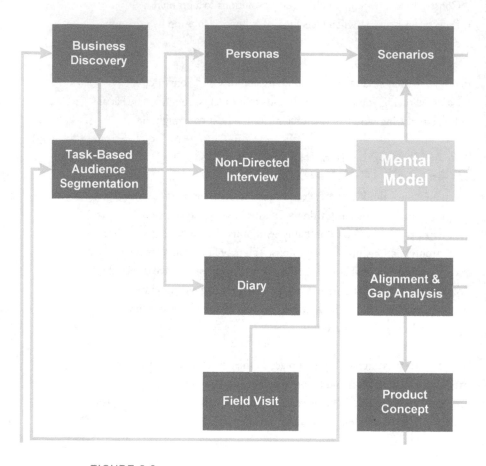

FIGURE 2.2.
Constellation of some user-centered design steps. (No wonder it seems so hard to figure out where to start!)

Mental models are also useful for things other than design. Sales and customer service can use the data to understand clientele better. MBAs and information designers can re-format the data into workflow and process diagrams. Project managers can use it to prioritize among a set of development options. I encourage you to reach out to these people and introduce them to any mental models that you create.

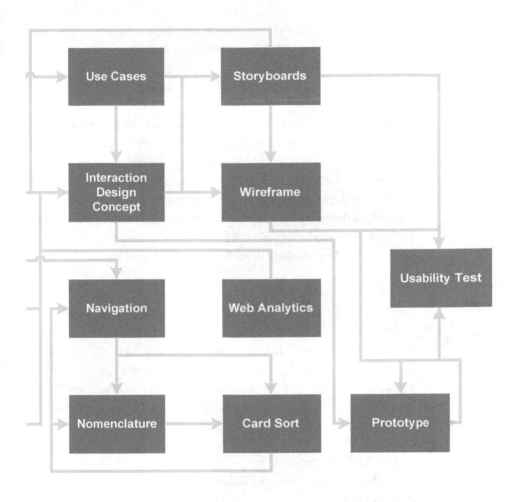

Within the realm of designing solutions, mental models provide a nexus for all the other tools in your toolbox. You can draw benefits from using mental models to support your personas and scenarios. Mental models along with web analytics and use cases influence your interaction design concepts. Prototypes coming out of these concepts undergo usability testing to touch base with the user.

There are a few additional techniques that could flow directly into or out of a mental model. I'll sketch these additional techniques here, and then dive into the main techniques later.

Input: Diaries

Diaries are a popular way to gather data in the user's voice. You could ask participants who are, for example, members of Weight Watchers to write down their daily successes and frustrations with the diet and exercise program they are trying to follow. You can comb through this data for behaviors and create a mental model from this analysis. Diaries do have a tendency to flit from subject to subject, however, without deep examination of a topic. This tendency can leave you with a spotty mental model. But if you have this data, go ahead and mine it for your mental model.

Input: Field Visits

Field visits conducted by a professional researcher produce a much deeper understanding, and all topics within a scope are likely to be covered. In this respect, field visit data is better for task analysis than diary content. To create a mental model of the participant's perspective, however, you will need to convert third-person notes into first-person behaviors. This translation is not insurmountable, and you will have a solid mental model as a result.

Output: Personas

According to Alan Cooper and Robert Reimann in *About Face 2.0: The Essentials of Interaction Design*, personas are user profiles that help the design team and other teams such as sales and marketing get more useful products into the hands of customers. If you review how they instruct you to write a persona, you will know to list experience goals, end goals, and life goals. The mental model focuses on the end goals (the things a person wishes to accomplish) and the life goals (the reasons why a person wishes to accomplish something—the larger picture). You can use this data to adjust your original task-based audience segments[5] and go on to build personas out of each segment with photos, names, and experience goals.

---

5 See Chapter 11, "Adjust the Audience Segments."

Output: Scenarios

A time-honored practice is writing scenarios that describe how a persona accomplishes a goal using a set of tools. Once you have a mental model, you can certainly write scenarios based on the meaningful tasks for your business.

I don't usually write scenarios, but it doesn't harm the design process if it is done correctly. Let me explain two things: Why I don't write them and what is incorrect execution. I don't write scenarios because I ordinarily work in quick-paced environments. I explore scenarios verbally with my team. To communicate these scenarios to other team members, we could use comics, storyboards, or videos,[6] but typically we don't have time. We go straight to prototypes, working directly and verbally with engineers.

What is an incorrect scenario? They are long, written descriptions that contain a lot of unnecessary detail, such as the specific buttons the user hits, every edge case,[7] or every error condition. Don't make it read too much like code, yet. Don't dwell on the facilities being used. Incorrect scenarios might also include non-relevant conditions that do not affect what the person is trying to accomplish, and might be better off as part of a persona description, such as gender or fashion preferences. In a situation where you're designing a clothing store, of course fashion preferences would affect how a customer behaves. But it's unlikely to affect how someone decides on a retirement plan.

Shortcuts and Other Ways to Use Mental Models

I have mentioned in the first chapter of this book that mental models can be applied to many different situations. Here I want to point out that

---

[6] A striking example was shown at MX—Managing Experience through Creative Leadership, San Francisco 2007, by Tim Brown, CEO of IDEO, in the form of a short video (really a series of still images with voice-over) of a futures trader making decisions about pork distribution for an upcoming sunny spring weekend. It was a story about how that trader used various tools, including phone, email, weather reports, and the prototype trading application to diagnose and act upon an opportunity.

[7] Wikipedia definition: "An edge case is a problem or situation that occurs only at an extreme (maximum or minimum) operating parameter." en.wikipedia.org/wiki/Edge_case

mental model research can scale to differently sized problems. Scalable adequacy is defined as "the effectiveness of a[n]...engineering...process when used on differently sized problems...Methods that omit unneeded notations and techniques without destroying overall functionality."[8] Mental models collect just enough data about users to help you determine where and how to concentrate your efforts. There are several ways to scale the mental model method.

When the Project is Almost/Already Finished

You might be at a late stage of design and development when you pick up this book. Don't despair; it's not too late to take advantage of a mental model, or at least a rough draft of one. Say you already have prototypes and usability test results. Say the users just don't get it. Sketching out a mental model will help you see exactly where your design veered off in a direction different than users. If the departure between your solution and user goals is significant, now is the time to convince someone to spend a little time on research so that the patches to the first version are not a waste of time. Developing something involves a lot of iteration, and if your first try is wide of the mark, subsequent tries will benefit more from a solid understanding.

What if your beta application is faring well and you want to know in which direction to move next? It's perfect timing to align functionality to a mental model and prioritize the gaps. When your team sees this diagram it will become a lot clearer how user research can help, even at this point.

In both of these instances you and your team can create a rough draft of a mental model in a matter of a day or two. First figure out which of your users you want to cover. Pore over existing user research data and try to extract knowledge using a behavior-and-philosophy-based perspective.[9] Then get in a room (virtual or physical) and brainstorm behaviors. From

---

8 "The Problem with Scalability," *Communications of the Association for Computing Machinery* (CACM), Sept 2000/Vol.43, No. 9 by Mauri Laitinen, Mohamed E. Fayad, and Robert P. Ward.

9 Existing user data may come in the form of preference or evaluative research. Try to deduce root causes, if they exist in the reports. (We'll cover root causes in great detail in Chapter 8.)

the existing research and your collective experience over the years, your team will be able to produce around 40% to 50% of the behaviors that actual research would create. Remind everyone to spit out descriptions from the user's point of view. Say, "There were a few people I heard from who did it this way," instead of, "I would have done it this way." Leave the personal pronoun "I" checked at the door. Think of real-life behaviors you have encountered, not your own real or hypothetical reactions.

After a few hours of brainstorming, take a break, then try grouping things together. You can create a reasonable draft of a mental model in just a few days. This method is how I did it during the heady dot-com boom at the end of the last century. Be aware, though, that every single mental model I produced with a team this way was missing at least one or more significant mental space. More ominously, only one of the dot-com companies I made a mental model for is still alive today,[10] and for them I employed the full-blown method with 34 interviews. Thought-provoking.

A draft mental model diagram can be the result of a few days worth of well-disciplined, task-oriented thinking on the part of the team. You can then check assumptions against this draft and even conduct gap analysis.

When You Have Little Time and Money

You might have extremely limited time or almost no budget. I have increasingly heard of teams following a three- or six-week "agile" development cycle. How does that leave you time to do proper user research?

Well, if it is going to happen, it has to occur in little chunks. Spend one of your development cycles mapping out the entire set of task-based audience segments you deal with, selecting the highest priority segment, and writing a recruiting screener to find these people. Hire a recruiter to line up some interview appointments. Then spend another development cycle interviewing four people from one of the audience segments (four

---

[10] The company that survived the bust under the guidance of CEO Peter Ostrow is www.testmart.com, selling previously owned, re-calibrated test and measurement equipment.

is the minimum to start seeing a pattern of repeated behaviors). Analyze the transcripts. At the end of this second cycle, you should have a mental model for that audience segment. At this point, you can do any of three things: You can proceed to another audience segment and interview those people; you can use the mental model you just created to design the solution you're working on; or you can take a step back and use the mental model to strategize where to focus your development efforts for the next few quarters. In any case, there are ways the process can be broken down to fit into your development cycles.

What if time is even tighter, and spending four weeks analyzing transcripts simply won't fit into your deadline? If you can, strive to conduct interviews with real people—the benefit of hearing their words is worth the cost of eating up a week or two of the time before your deadline. Instead of transcribing those interviews, my frequent collaborator Mary Piontkowski suggests capturing rough notes about behaviors in real time as you conduct the interviews, or creating these behaviors right after the interview from the notes you took. Without a transcript you will probably miss half of the behaviors, but the important ones will stand out in your mind and your notes. That will be good enough for a shortcut.

And what if there is no time to conduct interviews? Talking to real people is the most important part of creating the mental model. If your organization already conducts usability tests with some regularity, piggyback short interviews on top of each session. Ask the participant to stay with you for an hour, and spend half the time on the usability test and half of the time conducting a non-leading interview. At least this way you will get a chance to talk to real people.

Those in charge of the development cycle schedules usually see the advantages of an underlying base of research. Work on persuading them to set aside resources for this re-usable, long-lived information.

When You Don't Have Enough Influence

You might not be able to persuade anyone to follow this method. This is an extremely frustrating position to be in, and I empathize. Don't give up. You can pull together a rough draft of a mental model by yourself, simply by listing behaviors and grouping them, then laying them out in towers and mental spaces. You will have to work based on your accumulated understanding of customer aims, and you will want to write the behaviors from the customer's point of view. In the end, you will have a mental model that probably shows 30% to 40% of the mental spaces and towers. Treat the diagram as a rough draft, and use it to persuade others on the team to investigate further.

I have also heard of practitioners "flying under the radar" so to speak. They lay out all the steps to create a mental model, including task-based audience segmentation, interviews, and analysis, but they spread them out over the course of several months. When they have an unscheduled hour or a break from their assigned projects, they conduct an interview or analyze a transcript. In the end, they have a solid mental model to present at design meetings. I have heard this wins the respect of management and clears the way for subsequent user research. Be warned that this approach takes a very dedicated, determined personality, but that might be you!

Six Shortcuts to Mental Models

What you might have already guessed is that the approach I describe in each of the scenarios doesn't only apply to that scenario. Go ahead and choose any of the shortcuts that seem likely to work best for you.

- *Rough Sketch*: Sketch a rough draft yourself

- *Rough Draft*: Gather your team and create a mental model based on existing data and your collective understanding

- *Rough Notes*: Conduct the interviews, but skip the transcripts and pull behaviors from your notes of the conversation

- *Fly Under the Radar*: Conduct interviews and do your analysis as you can, over the course of several months

- *30-Day Cycles*: Go ahead and conduct interviews, but focus on just one or two audience segments, narrowing your data set from four to 10 interviews

- *Piggyback on Usability Tests*: After each half-hour usability session, tack on a half-hour interview

In the next chapter, I'll go into whom to include in your work and when to do it.

CHAPTER 3

Who?
Mental Model
Team Participants

Project Leader   40
Project Practitioners   41
Project Guides   42
Project Support   43

Whom should you include when choosing the members of your team during the mental model process? What kinds of skills do you need? Who will contribute and in what way? There are three basic types of participants: the project leader, the practitioners doing the research and analysis, and the project guides—stakeholders who review the work periodically, ensure it takes into account the details specific to each of their departments, and direct the focus of the project. This chapter outlines the ideal situation and encourages you to reach out to those with whom you might not normally work.

Project Leader

A researcher, information architect, interaction designer, or other person familiar with user research and design should lead the project. "Leading the project" means running all the workshops, staying on top of findings and shifting the direction of research accordingly, and actually conducting some of the work. This person should be completely familiar with the reasons for the research and the methods used. In other words, this person should be able to create and distribute the mental model herself. Think of this person as the one who either got the project approved or was appointed its shepherd by someone else who won the project. This person should also be prepared to lead the applied half of the project—that is, deriving architecture, analyzing gaps, leading project prioritization sessions, and mining the mental model for innovative ideas. There is value in having the same person play both the research and the architect/designer roles.

A product manager or project manager can assist with project tracking, logistics, team management, and political battles. Often, this role is filled by a person different from the project leader. Together, the two can coordinate the participating team members and the concepts generated.

This is a good place to say that the project leader will listen to all members of her team, no matter who they are. There's a saying, "It's hard to fly like an eagle when you surround yourself with turkeys." I wonder about the notion of ever thinking of your teammates as turkeys. People are smart.[1] There are plenty of books that can help you coax out their skills and give them confidence to do great work. A mental model will fall flat if the team creating it can't express ideas together and analyze concepts from different perspectives.

Project Practitioners

The people who actually conduct the research and create the mental model are responsible for the work of the project. Usually, one of the practitioners is the project leader. Practitioners will segment the audiences by task and create a screener for recruiting. They will draft the interview prompts and conduct the interviews. They will comb through the transcripts in search of behaviors and philosophies, and they will group the items they find by affinity. In short, they do all the work, plus show up at workshops with the project leader and the project guides to pick through all the details of what they've completed so far. It's not the easiest thing in the world.

Project practitioners can come from many different backgrounds. The only requirement that really seems to matter is a detail-oriented personality type. As long as everyone understands the process and can hang in there through the detail work, things will go smoothly. I have worked with people who have backgrounds that range from paralegal work to recreational administration. The method can be successful "even if the team members make a few mistakes along the way, and are not born information architects."[2] Your teammates just need to have a strong ability to understand new things, adapt, and think outside the structure they've already built in their minds.

---

[1] People that you're likely to be working with are smart, that is.

[2] Craig Duncan emailing about a recent mental model project he led for the Information Management Unit at the United Nations in Geneva, Switzerland.

Project Guides

A mental model project can have all sorts of intentional or unwitting guides. It is the project leader's job to hunt for as many different sources as she can. Start with the obvious: Directors and other executives in charge of fulfilling the mission of your organization will be able to tell you the long-term strategy they are following. They can tell you their objectives, predictions, and perceived challenges for the organization. They can talk about places where things went wrong in the past. Moreover, they can talk about the strengths and weaknesses of the internal organizations, and who makes the final decisions.

Reach out to people in customer service, marketing, and sales roles. These people are continuously in touch with users and have absorbed a lot of perspective and understanding during their interactions. Explore the ideas they've gathered. Bring in folks from technology. Find out what's on the horizon that they're excited about. Ask them about what initiatives they have faith in or don't see leading to success. Talk to the leaders of each product line and find out about their plans.

Invite a representative from each of these areas to be a member of your core team. The core team will meet several times during the project to review findings, ask questions, and keep the conceptual scope of the project on track. The more perspectives that participate, even at a light level, the better your understanding of where the mental model fits into all the things the organization is doing. In addition, it can be a great way to create widespread ownership of the design and architecture of your solution.

Project Support

There are a few more people to mention, namely the recruiters and the transcribers. Recruiters will find people for you to interview, and transcribers will type out recordings of these interviews, which are almost always conducted by phone. I highly recommend hiring professional agencies for both of these roles. If you can't find a good recruiting agency or if the decision was made to ask someone internal to your organization to recruit, you'll need someone with an outgoing personality who looks forward to interacting with many strangers each day. Many of us in the design industry are not the in-your-face cold-calling personality type.

Recruiting Makes Me Feel Rejected

"I have a hard time dealing with rejection in the first place."

—Deborah Nagai, Sybase

Finding a transcriber internally is more difficult. There may be a fast typist who is eager to help the project, but giving him 10 or 20 hours worth of interview recordings to transcribe is asking too much. It would be better to find 10 or 20 good typists each willing to transcribe one hour. It would be best, however, to just find the money in your budget to hire a professional. Speaking of budget, how much time and money does a mental model project require? Appendix A (see ⚑ www.rosenfeldmedia.com/books/ mental-models/content/appendix_a), outlines a range of costs, from quick-and-easy to full-blown.

The next six chapters will examine in depth how to create a mental model.

CHAPTER 4

Define Task-
Based Audience
Segments

Task-Based Audience Segments   46
Set Research Scope   65

Before you begin any kind of research, you need to decide whom to study. In traditional usability research and marketing studies, this has been determined by demographic or psychographic segmentation. Additionally, personality types like VALS[1] or Myers-Briggs[2] have made up the criteria by which research participants were chosen. While personality types do touch upon behavior,[3] generative research for building mental models requires that you select from groups of people who want to get different things done. Because you will want to tailor your end solutions to fit each audience exactly, grouping audiences by differences in behavior is important. You want to end up with solutions that match actions and philosophies closely rather than with one solution that fits several audiences loosely. Figure out what people want to accomplish, look for differences, and group accordingly. These are task-based audience segments.

Task-Based Audience Segments

Task-based audience segments are, quite simply, groups of people who do similar things. Task-based audience segments aren't about people who tend to select the same product—as in traditional marketing segments. Traditional marketing audience segments emphasize consumer tendencies because they were created to communicate a custom message to potential customers. People who have a similar outgoing personality or share an interest in wine tasting may buy the same magazine, but when it comes to handling their email inbox, they might act differently. You probably have been trained to think that "teens" and "senior citizens" fall into two

---

[1] VALS Personality traits that drive consumer behavior (formerly "Values and Lifestyle"), developed by SRI; www.sric-bi.com/VALS/

[2] Myers-Briggs psychological types describing information consumption, decision making, energy, and closure; www.myersbriggs.org/

[3] In 2001 Horacio D. Rozanski, Gerry Bollman, and Martin Lipman published an article in the Strategy + Business online magazine called "Seize the Occasion: Usage-Based Segmentation for Internet Marketers." The article urged researchers beyond demographics, but fell short of task-based audience segments, by using seven segments defined by internet behavior, such as session length, time per page, category concentration, and site familiarity. tinyurl.com/2c9cug

separate groups; however, when it comes to writing movie reviews, for example, they both may do the same thing. The teenage boy might email a movie review to all his friends. The 70-year-old woman might write film reviews for a local newspaper. For mental model research, these two belong to the same group.

You will need to drop the standard market aggregates. Instead, pay attention to the main ways in which people behave differently when engaged in what your product will solve. When you conduct research interviews, you will want to be sure that you elicit as many tasks, philosophies, and behaviors as possible, so that you can create a diagram that covers as much breadth as possible. If by chance you interview a teenager and a senior citizen, and you focus on the fact that they both write movie reviews and behave similarly with respect to seeing movies, you might not hear about the social moviegoer for whom movies represent more of a chance to see friends than a serious study of the film.

To avoid the traditional market segments trap, possibly creating a group wherein people do things differently or possibly missing a group entirely, I begin each project with a task-based audience segmentation exercise. The exercise has three steps:

1. **List Distinguishing Behaviors**. Sketch out all the ways many types of individuals might behave differently.
2. **Group the Behaviors**. Study these behaviors and put them into groups.
3. **Name the Groups**. Assign provisional labels to the groups.

There are simple cases and very complex cases. On the simple end of the spectrum, you might see groups like "engineer," "production director," or "sales rep." It's easy to see the difference in the roles these segments play. However, when you stray into more cultural or restricted scopes of research, such as "sending greeting cards" or "managing household telephone, internet, and cable services," it's harder to illuminate these differences. Because you can already recognize and create the easy groups, I will focus on the latter, more difficult scenarios.

Find a parking space	Write part of a screenplay
Run through the parking lot because I'm late	Act out a movie with friends
Boo and hiss at the villan	Act in character; make up my own adventures
Applaud the hero	Make my own costume
Scream at the scary parts	Buy a costume
Make out with my date	Become a favorite character for Halloween
Drive to the theater	Talk to my friends about my favorite characters
Take public transit to the theater	Talk to friends about my favorite actor/actress
Walk to the theater	Talk to my friends about my favorite director
Blink eyes to adjust to daylight upon exiting	Put up posters of my favorite movies
Watch all the credits	Collect posters of my favorite movies
Leave before the credits roll	Put up posters of my favorite actor/actress
Leave a film I don't like before the end	Collect fan memorabilia
Hide my eyes at the scary parts	Buy and make a model from a movie
Lift my feet from the floor at the scary parts	Collect models from movies (like Godzilla)
Clutch my friend's arm at the scary parts	Search for rare movie memorabilia
Tell my friend the ending I have guessed	Bid on rare movie memoriabilia
Go back for a refill on popcorn or soda	Attend movie/genre conferences (like Star Trek)
Spill soda/popcorn	Speak lines along with actors during movie
Leave container/wrapper on the floor	Sing songs from the movie
Put wrapper in the garbage on my way out	Get the movie soundtrack
Dream of meeting the leading man/lady	Buy other music from the composer
Buy action figures	Pick movie based on the composer
Buy the DVD/video when it comes out	Collect props from movies
Buy the t-shirt	Watch "the making of..."
Buy clothes like the characters wore	Watch actor interviews
Dress up in costume like the characters	Tune in when actors are interviewed on TV
Write a movie review	Watch director interviews
Tell my friends about the movie	Watch composer interviews
Read the book first	Make my own props
Read the book afterwards	Buy reproductions of movie props
Complain how the movie veered from book	Create a home theater
Exclaim how movie was adapted from book	Buy a surround sound system
Complain about the violence/sex	Buy a DVD Player/VCR
Buy products placed in the movie	Buy a better TV
Recite lines from a movie	Buy a projector and a screen
Pretend I'm the hero/villan	Rearrange furniture for best viewing
Pretend I'm the leading lady/man	Close curtains/dim lights
Learn a skill I saw in a movie	Collect cards based on the movie (Pokemon?)

FIGURE 4.1.

Brainstorm the different things people do before, while, and after seeing a movie.

Step 1: List Distinguishing Tasks

The first step is fairly simple. Thinking about our moviegoers, fill a document or a whiteboard with things people do when they go to the movies. At first, start with anything, like "buy tickets." It turns out that practically everyone buys tickets, so this particular action later gets tagged as "universal to all segments" and ignored. But it is a good task to get you going. List the activities using a verb-noun format. Make sure you think of tasks outside your own personal experience.[4] Keep brainstorming[5] new ideas until you have 150 or 200 of them (see Figure 4.1). This step takes an hour or two.

While generating this list, your team will hit lulls when you're sure you can't possibly think of anything else. If a lull lasts for more than a few minutes, point to an existing item and ask, "Is there more to it? What is at the root of this? Why are they doing it?" Usually another burst of ideas will occur.

Of course you can't be assured you'll think of absolutely everything a person does, but remember this exercise is only deciding whom to interview. The interview data itself will give you a more complete account of tasks and philosophies, and you can adjust these audience segments later.

At the end of this step, run through your list of tasks, and group duplicate and similar tasks. In the example shown in Figure 4.2, I grouped similar tasks and proceeded to work with only the tasks in bold. This step makes the list easier to handle, reducing it to about 75 items.

---

4 In fact, I recommend not using the personal pronoun "I" at all. Try to think of other people you know who go to movies and mention what you know about their behavior. "Katie chooses a movie by going to her favorite theater and seeing what's playing next." Then convert the task to the personal pronoun for the list.

5 June Cohen includes a good presentation of how to run a brainstorming session in her book *The Unusually Useful Web Book*, page 326.

Watch all the credits	**Go to movies alone**
Watch all the credits	Go to movies alone
Leave before the credits roll	**Dissect a film with friends**
Leave before the credits roll	Complain how movie veered from book
	Exclaim how movie was adapted from book
Leave a film I don't like before the end	Complain about the violence/sex
Leave a film I don't like before the end	Discuss movie with friends
Leave mess for janitors	**Tell everyone to see a film I enjoyed**
Spill soda/popcorn	Tell my friends about the movie
Leave container/wrapper on the floor	
	Share enthusiasm about movies
Throw away own garbage	Talk to my friends about my favorite characters
Put wrapper in the garbage on my way out	Talk to friends about my favorite actor/actress
	Talk to my friends about my favorite director
Blink eyes to adjust daylight upon exiting	
Blink eyes to adjust to daylight upon exiting	**Write a movie review**
	Write a movie review
See movies on opening night	
See most movies on opening night	**Go out with friends after a movie for food**
See a movie on opening night/at a gala	Go out with friends after a movie for food
Go to a film festival	**Listen to the soundtrack and related music**
Study the festival schedule	Get the movie soundtrack
Study the festival schedule with a friend	Buy other music from the composer
Choose schedule of films to watch at festival	
Attend a film festival	**Read the book**
	Read the book first
Watch movie multiple times in theater	Read the book afterwards
Watch movie multiple times in theater	
	Watch programs about the movies
Decide where/when to see a movie	Watch "the making of..."
Find out playing schedules	Watch actor interviews
Pick a theater nearby	Tune in when actors are interviewed on TV
Check friends' schedules	Watch director interviews
	Watch composer interviews
Go to movie with friends	
Meet friends at the theater	**Recite/sing from the movie**
Go to movies with co-workers	Recite lines from a movie
	Sing songs from the movie

FIGURE 4.2.

Eliminate or group duplicate and similar tasks.

Step 2: Group the Tasks

The second step is to group these items where there is *behavior affinity*. In other words, think about the people who do these things and group them into types, such as "people who study films" and "people who go to movies to be with friends." It is important not to group by *verb affinity* (such as "choose a movie" or "go to the theater"), since that comes later when you build the mental model. *At this point you want to identify actors, not actions.*

How to Segment Complicated Audiences

In audiences that are difficult to segment, such as "people looking for a date," it's easy to get confused. It would be a mistake to make audience segments like "Advertisers of Single Status," "Date Arrangers," and "Date Analyzers," because one person could do all of these things. Those are actually mental spaces: "Let People Know That I'm Available," "Arrange for a Date," and "Analyze My Date Afterwards with Friend." What you should focus on is the differentiating behavior of people, such as "people who are outgoing and feel confident that love is possible" or "people who feel shy or unlovable." The tasks that differentiate these groups are things like "go up and ask someone for a date" versus "daydream about a person I like." A task like "decide which restaurant to eat at" might apply to both groups, so it's universal and you can ignore it. Focus on the distinguishing tasks.

To make sure that you aren't missing a particular audience segment, my colleague Mary Piontkowski suggests you may want to list some probable mental spaces and cross check them with your groups.

This second step, grouping the tasks into behavioral affinity groups, gets a little more complex. If you're lucky or gifted or have easy-to-see differentiated task groups, you can do this with one pass through the tasks. For one client who was studying the dating world, we were lucky. There were obvious differences between the segment "who felt love was everywhere" and the segment "who had difficulty finding love." The former did a lot of connecting, and the latter did a lot of worrying and discussing.

Usually it's not that easy. I often take two passes—one pass during which I sketch out some potential performers of each task, and another pass during which I see how the tasks and performers fall into groups. For the moviegoers, I started going down the list of tasks and writing down the name of a performer who might do this task. Several would occur to me for each task. Soon, the same performers were suggesting themselves for subsequent tasks. I made a matrix like that shown in Figure 4.3, where I listed the tasks down the left and the performers across the top, then I put an "x" in each cell that applied. For "Choose film based on actor/actress" I listed "Actor/Actress Fan," "Director/Composer Fan" because it's similar, "Submerse in Another World" for the people who wanted to pretend they were on screen with the actress, and "Serious Collector" for someone who wants to own every Ridley Scott movie. Yes, the names of these performers can be pretty subjective and judgmental, but that's okay for now. These are not the names of our audience segments, yet.

Next is an exercise in pattern-matching. I ignore the text in the left column and look at the patterns of x's in the matrix. Which horizontal rows have similar patterns? "Dissect a film with friends" has x's from the third column through the seventh column, three x's at the ninth column. If you look down the list, you see "Write a movie review" also has this pattern, although without the x's in the ninth column. Group these two rows together for now, in a separate place on the spreadsheet. Next, look at the similarity between "Put together a model from a movie" and "Trade memorabilia with friends." Both rows have the same pattern of x's. Group these two rows together farther down on the spreadsheet, separate from the two you've already grouped there, since the patterns of x's are different from one another. Keep looking for patterns. You needn't be exact. (Note that if you have rows with x's almost all the way across, then delete them. These rows represent universal tasks and, thus, can be ignored. You are interested in distinctive tasks—those tasks that may separate one set of people from another.) In Figure 4.4, you can see that I used a colored background under the x's to show the larger pattern. This will help you more easily see the patterns so you can move one row to be with another similar row.

	Child with Parent	Youth with Friends	Submerse in Another World	Movie Snob	Technical Moviegoer	Genre Fan	Director/Composer Fan	Actor/Actress Fan	Dating	Group Entertainment Organizer
Choose film based on actor/actress			x				x	x		
Go to movies alone				x	x		x			
Dissect a film with friends			x	x	x	x	x		x	x
Tell everyone to see a film I enjoyed		x	x	x			x	x		x
Share enthusiasm about movies	x	x	x				x	x		x
Write a movie review			x	x	x	x	x			
Go out with friends after a movie for food		x	x				x		x	x
Listen to the soundtrack and related music	x	x	x	x	x	x				
Read the book	x		x	x						
Watch programs about the movies			x	x	x		x	x		
Recite/sing from the movie	x	x	x				x	x		
Make preparations to re-enact the movie	x	x	x							
Re-enact the movie	x	x	x							
Act like the characters in play	x	x	x							
Act like the characters in normal life		x	x							
Wear a costume to a film event		x	x				x	x		
Wear a costume to an unrelated event	x	x	x							
Dream of being the character in the movie	x	x	x							
Attend movie/genre conference			x							
Create my own movie		x	x	x	x					
Collect memorabilia	x	x	x				x	x		
Collect actual props and wardrobe			x				x	x		
Put together a model from a movie	x	x	x				x	x		
Trade memorabilia with friends	x	x	x				x	x		
Put up posters	x	x	x						x	

FIGURE 4.3.
Draft some performers for each task.

You can see that the patterns are not exact—there are outlying x's. But there are different green chunks that each row has, and there are differences among the sets of these green chunks.

Review these sets and convince yourself that the tasks assigned to each group really do make sense together. Most likely you will want to move a row or two to be with another set, and it is always gratifying to see how the x's align better with the new set.

	Child with Parent	Youth with Friends	Submerse in Another World	Movie Snob	Technical Moviegoer	Genre Fan	Director/Composer Fan	Actor/Actress Fan	Dating	Group Entertainment Organizer
Put together a model from a movie	x	x	x		x		x			
Trade memorabilia with friends	x	x	x		x		x			
Put up posters	x	x	x					x		
Collect memorabilia	x	x	x		x		x	x		
Watch the movie multiple times in the theater		x	x		x	x	x	x		
Surmount difficulty to get to seat	x	x	x				x	x		x
Sneak in to theater		x								
Choose seat for immersion		x	x		x	x		x		
Choose film based on actor/actress			x					x		
Go to movies alone				x	x		x			
Throw away own garbage				x	x		x			
Leave a film I don't like before the end	x			x						
React soundlessly	x			x			x	x		
Choose seat for privacy				x	x	x	x	x	x	
Skip/Postpone movies I might not like				x	x	x	x	x		
Create a home theater			x	x	x		x			
Collect editions of films			x	x	x		x			
Watch programs about the movies			x	x	x		x	x		
Buy the movie for repeat viewing			x	x	x	x	x			
Dissect a film with friends			x	x	x	x	x		x	x
Write a movie review			x	x	x	x	x			
Watch all the credits			x	x	x	x	x		x	
Avoid leaving in middle of film			x	x	x	x	x		x	

FIGURE 4.4.
Rows of x's with similar patterns where I've grouped them farther down the spreadsheet.

Step 3: Name the Groups

In the final step, you simply name the different sets of green chunks. Then you copy the tasks and performer names to a separate page, and the team sets about brainstorming a name. The rightmost column in Figure 4.5 shows a list of names we brainstormed for the moviegoer sets. The first two columns list the tasks and performers as reminders to help come

Primary Tasks	Performers	Brainstorm Audience Segment Name
Social Movie Goer		
React out loud to the movie	Child with Parent	Entertainment Enthusiast
Participate in movie	Child with Friends	Social Climber
Tell everyone to see film I enjoyed	Submerse in Another World	Raging Hormones
Share enthusiasm about movies	Genre Fan	Social Movie-Goer
Go out after movie for food	Actor/Actress Fan	Just Entertain Me
Go to movie with friends	Dating	Blockbuster Fan
Leave mess for janitors	Group Entertainment Organizer	Experience Immersion
Indulge in refreshments	Gain Social Acceptance	Audience Immersion Fan
	Indulging Urges	Group Experience
		Pop Culture Fan
Movie Buff		
Get to theater on time	Submerse in Another World	
Buy tickets ahead of time	Movie Snob	
Make sacrifice to get tickets	Technical Moviegoer	Serious Movie Buff
Stand in line for the movie	Genre Fan	Methodical
Make sacrifice to get good seats	Director/Composer Fan	Selective
Watch movie at home	Friend Bringing Friend	Film Aficionado
Choose film based on genre	Dating	Movie Expert
See movies on opening night	Group Entertainment Organizer	An Authority
Go to a film festival	Gain Social Acceptance	Watches Everything
Cut school/work to see a movie	Serious Collector	Knowledgeable
	Book Reader	Movie Fan
Big Fan		
Put together a model from a movie		Fan Club
Trade memorabilia with friends	Child with Parent	Focused on One Thing
Put up posters	Child with Friends	Genre Aficionado
Collect memorabilia	Submerse in Another World	Fan of the Director
Watch movie multiple times	Technical Moviegoer	Celebrity Spotter
Surmount difficulty to get to seat	Actor/Actress Fan	Rub Shoulders w/Famous
Sneak in to theater	Serious Collector	Belong to the Clique
Choose film based on actor/actress		Elitist
Choose seat for immersion		Technical Buff

FIGURE 4.5.
Brainstorm a likely name for the set, based on its tasks and performers.

up with names. After you have a satisfying list of potential segment names, you can vote on the best name. The winning names are listed at the top of each group in the horizontal light green heading.

Often I will test these names by describing them to people and asking if they can tell me to which segment they belong. This is not necessary if your segments are pretty well understood, such as "Truck Driver," "Mechanic," and "Dispatcher." For example, with the matchmaking company, we ended up with audience segments "Get on the Love Train" and "Trying Too Hard," plus two others called "See What Happens" and "Think It Through." It was pretty easy for people to figure out where they belonged. But the moviegoer segments were unusual. Indeed, when I asked people which group they thought they fell into, several were perplexed as to the differences among the Movie Buff, the Big Fan, and the Film Purist. Also the Social Moviegoer label felt too restrictive to people. This red flag made me take a second look at the segments.

First, here are the descriptions that I told various people, leaving out the universal things like "choose a comfortable seat," "prefer clean theaters," or " buy tickets."

- **Social Moviegoer**. This person enjoys films more for the pleasure of being with friends than anything. She will pick (or agree to) whichever film appeals to everyone. She usually goes to a theater she knows that has easy parking, no crowds, no waiting, and no lines for tickets. She will interact with her friends during the movie with eye contact or comments. She is likely to react aloud to certain scenes. She will spend a few hours with her friends right before or after the movie, and usually dinner or a café is involved. After the movie, she likes to talk about the films with friends to find out what interests they have in common, what she did and didn't like about the movie, and explore suggestions about how the film could have been different.

- **Movie Buff**. This person enjoys all sorts of films, all sorts of styles, all sorts of actors and directors. The art of the storytelling is what interests her. She wants to see the craft and quality of the film for its aesthetic sake. This person studies the film. She may have taken formal classes about filmmaking and most likely tracks what is being developed, release dates, critics' reviews, and box office success. She may be interested in non-commercial, movie-related art. She is likely to see films on opening night because the print is fresh. She will be on the lookout for inside jokes. She likes to watch her companions' reactions during the film. She will choose the appropriate theater for the film: big crowds and group experience for blockbusters or horror films, small theaters for drama, or new theaters for films with extravagant special effects. She will buy advance tickets if she expects a crowd. She may take a friend to the movie (second time for her), so he can enjoy it. She will go to film festivals. She may write film reviews for friends. She may screen films for friends or start writing a screenplay of her own.

- **Big Fan**. This person is really excited about a particular film genre/actor/director. She will track the release date. She will read the book to prepare herself for the experience. She will make a sacrifice to get tickets and a good seat. She is likely to see films on opening night because the crowd is more energetic. When watching the film, she wants total audio/visual immersion. She likes to watch her companions' reactions during the film. She is likely to see the film multiple times while it is playing in theaters. She will go to film festivals. Afterwards, she is likely to quote from the movie, sing songs from the movie, and listen to the soundtrack. She will buy memorabilia of the film, such as posters, action figures, or props, and she may trade these items with other fans. She will buy a copy of the film for repeat viewing. She may write film reviews for friends.

- **Film Purist**. This person carefully chooses films that will enhance her life. She may choose a film because it will educate her, show her a different perspective, or explore unfamiliar cultures; it will depict far-away places, or certain emotions, for example. She is after an intellectual, emotional, or perception-changing experience, and usually this is not mainstream films. She may look up additional information at home afterwards. She will choose high-quality films, either in terms of the directing, the acting, the score, the cinematography, the choreography, and so forth. She might instead just go to a theater she trusts and attend the film that is playing soonest. She is interested in what the director has to say and the reason he created the film, and she will sit and respectfully observe the artists' work. She wants to get the full experience the director wanted the audience to have. She avoids crowds. She prefers to see a film without distraction from others and will not react aloud in the theater, but will cry, etc. She will try to let the information/feeling linger. This person is likely to see films alone.

- **Make-Believe Artist**. This person not only wants to suspend disbelief, but wants to be a part of the world illustrated in the movie. She is more likely to wear a costume like those worn in the movie, or speak and behave like her favorite character(s). Back home, she may re-enact scenes from the film or pull elements from that world into her own. She may read the book to experience the film's world in more depth. She is likely to choose films with familiar characters. (Imagine this person as a six-year old dressing like a Disney princess, or a Trekkie dressing as a Klingon.)

- **Enabler**. This person is going to the movie purely to make someone else happy. Perhaps she is taking a dependent who has chosen the film, or perhaps she is there to make someone else happy or to avoid friction. Perhaps she is there to make an impression on another person. This person is likely to leave the theater at the end of the film, before the credits roll, as well as leave temporarily during the film to use the restroom or get refreshments. Most likely this person is more focused on the other(s) she is with than on the film itself. (Imagine anyone from a mother with her young kids to a guy with a date.)

Of the descriptions above, people could not always figure out where they landed. True, sometimes people move between roles, but usually they would be able to identify primarily with one audience segment. For example, Carolyn is the mother of three children and has in-laws living with her. She often takes her children to the movies. Her mother-in-law enjoys romances, so Carolyn takes her to those films as well. She buys a lot of DVDs for the kids at home. She watches movies at home that her husband picks about three times a month. So far, Carolyn is an Enabler.

However, she escapes to the theater with her female friends about once a year. The movie they choose doesn't really matter. This part of her behavior makes her a Social Moviegoer.

How to Tell if Existing Segments Need Adjusting

There are at least two tests to determine if your audience segments are close-to-right.

- Describe the segments to someone and ask them which group they fall into. Do this six or seven times, after which you'll be able to tell whether your groups are appropriate.

- Check to see if you've mistakenly named the groups after actions rather than performers. If your names are closer to "Chopper," "Stirrer," and "Plate Arranger" for people who work in a commercial kitchen, you might want to re-assess. In this case, the traditional difference between the things a chef de cuisine must accomplish versus what a sous chef or a pastry chef does is probably closer to what you want. The latter two chop and stir and cook, but the chef de cuisine has the responsibility for day-to-day operations in the kitchen, including calling out orders to the sous chef, the pastry chef, and anyone who is a chef of the line so that they can put out the food on time.[*]

---

[*] Thanks to Ryan Freitas for a peek into the "back of the house," meaning the kitchen side of a restaurant. See his writing at secondverse.wordpress.com

---

So I started looking at the root tasks for our descriptions and realized something. There are three core attributes that define differences in people's behavior with respect to movies: their interest in the *story*, their appreciation of the *craft* of moviemaking, and what kind of *companionship* they choose when seeing movies. Each of these three attributes is represented by a continuum of intensity, from low to high. Of the three, a person's position on the continuum of Story tends to vary depending on the film, but the others remain somewhat constant. All seem to shift with maturity, though.

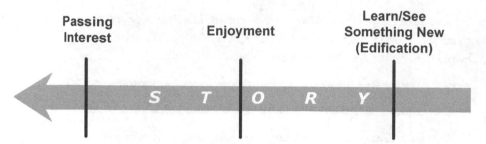

Passing Interest

Enjoyment

Learn/See Something New (Edification)

S T O R Y

This person chooses the film that is most likely to entertain her from a given set of films. Most likely the fact she is going to see a movie is pre-determined, and she merely selects from the stories available.

This person has probably heard of the film, decided a while ago to see it, looks forward to the experience, and feels pleased with her choice. The film successfully entertains her. She may talk about the film with friends afterwards to compare notes about the story and characters.

(Film Purist) This person carefully chooses films that will educate her, show her a different perspective, explore unfamiliar cultures, show far-away places, certain emotions, etc. She is after an intellectual or perception-changing experience, and usually this is not mainstream films. She may look up additional information at home afterwards. She wants to get the full experience the director wanted the audience to have. She avoids crowds. She will not react out loud in the theater. She will try to let the feeling linger.

FIGURE 4.6.
Continuum for Story.

I came up with the following drawings, using a lot of the material I had already gleaned from the brainstorming exercise with my team (Figures 4.6 through 4.8). It turned out that three of my original segments emphasized Story, one of them was an instance of Craft, and two focused on Companionship. I have added six additional descriptions where there were gaps in the continuum for each.

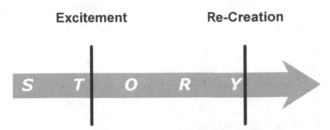

Excitement   **Re-Creation**

S T O R Y

(Big Fan) This person will make a sacrifice to get tickets and a good seat. She is likely to see films on opening night because the crowd is more energetic. When watching the film, she wants total audio/visual immersion. She feels a strong connection to the story. Afterwards, she is likely to quote from the movie, sing songs from the movie, and listen to the soundtrack. She will buy memorabilia of the film, such as posters, action figures, and props. She will buy a copy of the film for repeat viewing.

(Make Believe Artist) Wants to be a part of the world illustrated in the movie. She is likely to wear a costume like those worn in the movie, or speak and behave like her favorite character(s). Back home, she may re-enact scenes from the film or pull elements from that world into her own. She may read the book to experience the film's world in more depth. She is likely to choose films with familiar characters.

Given these three scales, my informal tests proved people could more easily pinpoint where they would position themselves. What remained was choosing which combinations we were interested in for our research.

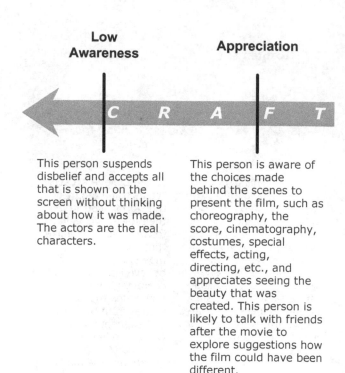

This person suspends disbelief and accepts all that is shown on the screen without thinking about how it was made. The actors are the real characters.

This person is aware of the choices made behind the scenes to present the film, such as choreography, the score, cinematography, costumes, special effects, acting, directing, etc., and appreciates seeing the beauty that was created. This person is likely to talk with friends after the movie to explore suggestions how the film could have been different.

FIGURE 4.7.
Continuum for Craft.

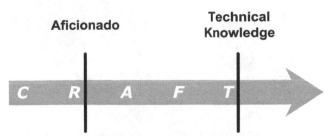

Aficionado	Technical Knowledge

This person has an interest in an actor, director, or another figure creating the film. Certain star actors may attract her attention, since she would want to see all of their work. She may be interested in what the director has to say and the reason he created the film, and she will sit and respectfully observe the artists' work. She will look for in-jokes. She will attend Film Festivals.

(Movie Buff) This person has knowledge of how films are created. This person studies films. She may have taken formal classes about film making and most likely tracks what's being developed, release dates, critics reviews, and box office success. She observes different directing styles and understands how the screenplay was written. She may write film reviews for her friends.

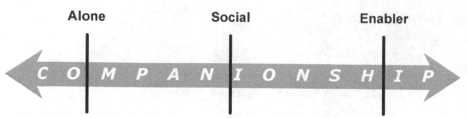

Alone

Social

Enabler

This person avoids crowds. She prefers to see a film without distraction from others, and will not react out loud in the theater.

(Social Movie-goer) This person enjoys being with friends at films. She will pick (or agree to) whichever film appeals to everyone. She usually goes to a theater she knows, has easy parking, no crowds, no waiting, and no lines for tickets. She likes to watch her companions' reactions during the film. She will interact with her friends during the movie with eye contact or comments. She will spend a few hours with her friends right before or after the movie, and usually dinner or a cafe is involved. She likes to talk about the films afterwards with friends to find out what interests they have in common or what she liked or not.

(Enabler) This person is going to the movie purely to make someone else happy. Perhaps she is taking a dependent who has chosen the film, or perhaps she is there at the bequest of someone else, to make them happy or avoid friction. Perhaps she is there to make an impression on another person. This person is likely to leave the theater at the end of the film, before the credits roll, as well as leave temporarily during the film, to use the restroom or get refreshments. Most likely this person is more focused on the other(s) she is with than on the film itself.

FIGURE 4.8.

Continuum for Companionship.

Set Research Scope

At this point you know all the possibilities you can recruit for your research. You may wish to examine whether all of these segments are important to your business. Lou Rosenfeld says, "This is one of those murky areas where information architecture and management responsibilities blur. Decision-makers might already have clear metrics in place for guiding such decisions. Conversely, the business may not have a good set of metrics and goals in place, complicating the definition of importance. In these situations, information architects and other UX people are playing a greater role in driving this discussion forward. We have no choice here, because we need the answers to do our work."[6] Selecting the top few segments to study from the entire set helps you scope your work. Paring down the number of segments also greatly reduces the cost and time required to create a mental model.

In the moviegoer example, there are 60 permutations to choose from. Furthermore, the imaginary film-distribution company I was doing the research for, JMS Entertainment, did not believe all aspects of Story-Craft-Companionship were big contributors to its bottom line. In a real scenario, I would sit down with the stakeholders and review each potential group, setting out the importance of that group to the business and weighing the importance of including them in the research. With my make-believe client, we selected the following circled areas in each spectrum that influenced the business strategy (Figures 4.9 and 4.10).

---

[6] Lou Rosenfeld's blog called *"Bloug,"* entry from April 4, 2007, *"The No-Knead Approach to Information Architecture (#3 of 5),"* tinyurl.com/yt9kcd

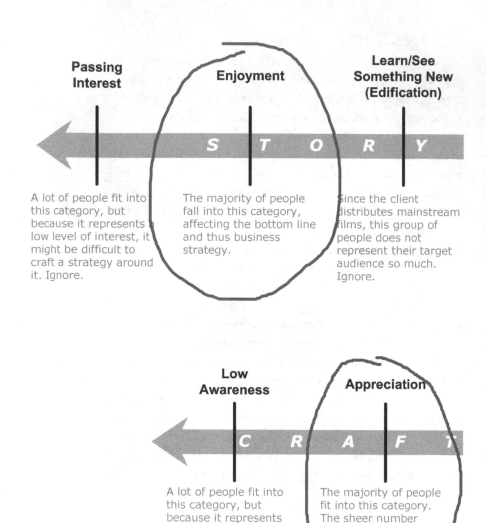

Passing Interest

Enjoyment

Learn/See Something New (Edification)

S T O R Y

A lot of people fit into this category, but because it represents a low level of interest, it might be difficult to craft a strategy around it. Ignore.

The majority of people fall into this category, affecting the bottom line and thus business strategy.

Since the client distributes mainstream films, this group of people does not represent their target audience so much. Ignore.

Low Awareness

Appreciation

C R A F T

A lot of people fit into this category, but because it represents lack of awareness, it might be difficult to craft a strategy around it. Ignore.

The majority of people fit into this category. The sheer number affects income and therefore business strategy.

FIGURE 4.9.
Work with stakeholders to determine a subset of the audience segments to study immediately, based on their importance to the business.

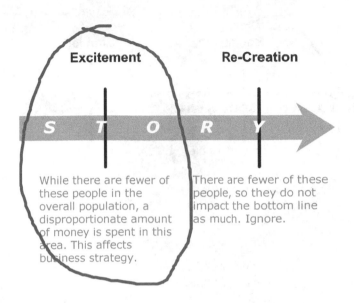

Excitement

Re-Creation

S T O R Y

While there are fewer of these people in the overall population, a disproportionate amount of money is spent in this area. This affects business strategy.

There are fewer of these people, so they do not impact the bottom line as much. Ignore.

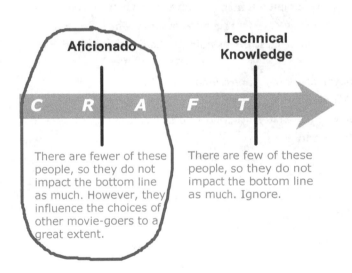

Aficionado

Technical Knowledge

C R A F T

There are fewer of these people, so they do not impact the bottom line as much. However, they influence the choices of other movie-goers to a great extent.

There are few of these people, so they do not impact the bottom line as much. Ignore.

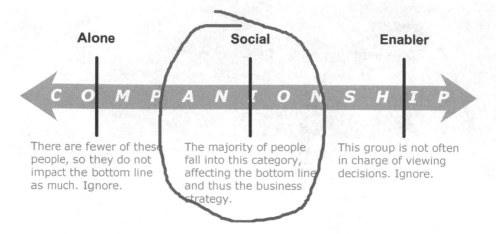

Alone

Social

Enabler

C O M P A N I O N S H I P

There are fewer of these people, so they do not impact the bottom line as much. Ignore.

The majority of people fall into this category, affecting the bottom line and thus the business strategy.

This group is not often in charge of viewing decisions. Ignore.

FIGURE 4.10.
Work with stakeholders to determine a subset of the audience segments to study immediately, based on their importance to the business.

With two segments of interest from Story, two from Craft, and one from Companionship, I now have four permutations from which to recruit. I created new segment names corresponding to these four combinations.

1. Story Enjoyment + Craft Appreciation + Social Companionship = **Typical Moviegoer**
2. Story Enjoyment + Craft Aficionado + Social Companionship = **Craft Aficionado**
3. Story Excitement + Craft Appreciation + Social Companionship = **Story Fanatic**
4. Story Excitement + Craft Aficionado + Social Companionship = **Total Film Buff**

Remember that these are working names for your segments. After you conduct the interviews and create the mental model, you will revisit these segments and adjust them based on your research. These task-based segments can be developed into personas,[7] famously introduced by Alan Cooper in his book, *The Inmates Are Running the Asylum*.

For now, we will leave the other 56 permutations out of our research. As time passes and more opportunities to do research present themselves, we can choose to explore other appropriate segments.

The next chapter outlines how to communicate to recruiters the number of people, with detailed selection criteria, you wish to interview.

---

7 Also see the note in Chapter 10 for more information on personas.

CHAPTER 5

Specify Recruiting Details

Estimate the Tally   72
Write the Screener   76
Coordinate Schedules   83
Recruit Participants   87

You know which audience segments you are interested in, but who should these people be, exactly, and how many of them? In this chapter you will start a new file, called the recruiting spreadsheet, which will specify the following details:

- How many of each audience segment to select
- Demographics to select, if applicable
- A qualification questionnaire for recruiters to use
- Questions to make certain that the candidate is able to carry on an hour-long conversation
- A schedule with available interview appointments
- A list of qualifying candidates

I will also outline a few recruiting approaches that I have had success with.

Estimate the Tally

Based on experience interviewing people over the past decade, I have concluded that you will hear the same sorts of tasks and philosophies after interviewing the fourth person from a particular segment. So, obviously, I recommend interviewing at least four people from each segment. During consecutive interviews, I pay attention to whether I am hearing the same things from each new person; if I'm not, I ask for one or two additional recruits from a particular group.

For the moviegoer research, to continue with our example, I needed a minimum of 16 participants, four from each group.

When One Person Represents Multiple Audience Segments

In some rare cases in which one person can actually be a member of two audience segments, you can interview one person about both aspects and count it as two audience segment interviews. While rare, this situation occurred with the information management and mobility company Sybase. We had three audience segments: Evaluator, Implementer, and Maintainer. Any one person at a company using Sybase products could play one or two roles. There were people who were Evaluator/Implementers, people who were Implementer/Maintainers, and people who were Evaluator/Maintainers. There were even two people who played all three roles. When counting up whom we interviewed, we counted by role rather than by person.

Before I recruited the moviegoer participants, though, I checked in with the JMS Entertainment stakeholders about other characteristics that were important to the business. Does gender, age, geographic location, income level, a college degree, or any of the typical demographic information make a difference whom we recruit? For instance, maybe there is a skew towards a male audience in most of the films they distribute, in which case it may help us to understand males more than females. It would certainly help to talk to people who see movies frequently—as opposed to accidentally recruiting someone who sees a movie only once or twice a year. The stakeholders also wanted to be sure there were no differences between geographic locations. They distribute movies in the United States and Canada, so an even distribution of participants would answer the question as to whether there were regional differences. To summarize the participants we were looking for, I set up the matrix shown in Figure 5.1 in a new file that I called the recruiting spreadsheet. This recruiting spreadsheet is available in the Resources section of the book site

🐘 www.rosenfeldmedia.com/books/mental-models/content/resources

	Recommended	Recommended Total
Audience Segments		
Typical Moviegoer (story enjoyment, craft appreciation)	4	16
Craft Aficionado (story enjoyment, craft aficionado)	4	
Story Fanatic (story excitement, craft appreciation)	4	
Total Film Buff (story excitement, craft aficionado)	4	
How many movies per month?		
Three or more movies	16	
Less than three movies	0	
Region		
West, Northwest, Southwest US	2	
Central US	2	
Northeastern US	2	
Southeast US	2	
Western Canada	2	
Central Canada	2	
Atlantic Seaboard Canada	2	
French-Speaking Canada	2	
Gender		
Female	5	16
Male	11	

FIGURE 5.1.

Audience segment and demographic goals for moviegoer recruiting.

In another example, I set up the demographic goals for the matchmaking company (Figure 5.2). In this case, there were more demographics that were important to the stakeholders. In addition to gender and a relatively even geographic distribution, stakeholders wanted me to interview people who have a college education, who have been out of college in the real world experiencing dating for a few years, and who have a mature dating goal. Moreover, they wanted me to interview people who were already taken (in a relationship), as well as those still looking.

Desired Number of Recruits by:		
Audience Segment		
The Good Samaritan	3	
Will Participate If Asked	2	
Social Connector	3	14
Self Conscious	3	
Expand the Possiblities	3	
Location		
Northwest Urban Area (Seattle)	2	
Southwest Urban Area (San Francisco Bay Area)	3	
North Central Urban Area (Chicago)	2	14
South Central Urban Area (Dallas)	2	
Northeast Urban Area (NYC)	3	
Southeast Urban Area (Atlanta)	2	
Gender		
Female	9	14
Male	5	
Age		
Out of College	14	14
In High School or College	0	
Dating Experience		
Two to Five Years After College	5	14
Five to Ten Years After College	5	
Ten Years or More After College	4	
Status		
Dating	10	14
Taken	4	
(If Single) Dating Goal		
Marriage Soon	7	10
Marriage Probably Down the Road	3	
Casual Dating	0	
Education		
No College Education	0	14
College Degree or Above	14	

FIGURE 5.2.
Demographics for recruits for matchmaking site Engage.com.

Be careful not to over-specify your demographic requirements. Only select what you suspect will make a difference in behavior—in addition to what behavior your audience segments define. Try to rely on your audience segments themselves for the greatest distinctions. For example, a company I worked with was sure there was a difference in the way people interacted with Human Resources based on their level of experience with the web. I pointed out that web experience had nothing to do with finding out your vacation balance, so we dropped the demographic.

Write the Screener

The next step for the moviegoer research was to design a screener for recruiting candidates. A screener is a list of questions the potential interviewee answers that will either qualify or disqualify her from participating in the research. I usually put the screener on a different tab in the same spreadsheet that contains the goals from Figure 5.1. If I am writing a screener for use by a recruiting service, I write it with the junior intern in mind—the person with the least experience who might be asked to phone potential participants. If I am writing a screener for use as a pop-up on a web site, I write it with the end-user in mind, naturally. In either case, I write the screener in survey format.

For the moviegoer research, I wrote the screener for a recruiting agency. I began the screener with a paragraph the recruiter could recite into the phone upon contacting someone (Figure 5.3).

> Hello, I represent a research firm that will give you a $50 gift card to participate in a study about movie-going. We are recruiting people to participate in one hour telephone interviews. If this interests you, I have a few qualifying questions to ask. The following questions help us find participants that match profiles we are interested in.

FIGURE 5.3.
Beginning of the moviegoer screener, written for a recruiting agency

Then I set up a series of screening questions with a column per candidate for their answers (as shown in Figure 5.4). I write all the questions to ask, one per row, in the first column. The subsequent 16 columns, each labeled with the participant identification numbers, contain the answers for each candidate. The recruiter can select the first unfilled column and move down the rows, asking questions of the potential recruit on the phone, and then marking answers in the column. I use 1's in the spreadsheet so that I can total them in the rightmost columns. You can also write a "count" formula to sum the number of x's in the cells, if you prefer to use x's instead of 1's.

		Participant		
	101	**102**	**103**	**104**
How many movies do you watch per month?				
Three or more movies	1	1	1	1
Less than three movies				
In what part of the country do you live?				
West, Northwest, Southwest US	1			
Central US				
Northeastern US			1	
Southeast US				
Western Canada		1		
Central Canada				1
Atlantic Seaboard Canada				
French-Speaking Canada				
And you are _____?				
Female	1		1	
Male		1		1

FIGURE 5.4.

Recruiters mark a "1" in the appropriate rows for each candidate they call. This example shows the completed form for all 16 candidates.

Because our moviegoer segments are not obvious, I needed to help the recruiters determine to which of the four segments a person belonged (Figure 5.5). For each of the three aspects, Story, Craft, and Companionship, I list a few defining details for each possible answer. The recruiter can read these aloud to the candidate, or ask the candidate to describe herself and choose the most closely related option.

Participant											
105	106	107	108	109	110	111	112	113	114	115	116
1	1	1	1	1	1	1	1	1	1	1	1
	1										
										1	1
							1				
1		1									
						1					
								1			
				1	1						
			1						1		
						1		1			1
1	1	1	1	1	1		1		1	1	

How do you typically approach the story and characters of a movie?
Choose the film that is most likely to distract me. (Passing Interest)
Look forward to seeing it; believe I will enjoy it. (Enjoyment)
Looking for an intellectual or perception-changing experience; may look up additional information at home. (Edification)
Prepare for the experience; make a sacrifice to get ticket or a good seat; go on opening night; likely to see the film multiple times. (Excitement)
Wear a costume like those in movie; speak and behave like favorite character. (Make Believe)

In general, how much do you notice the craft of how a movie is made?
Accept all that is shown on the screen without thinking about how it was made; actors are the real characters. (Low Awareness)
Aware of the choices made behind the scenes; appreciate seeing the beauty that was created. (Appreciation)
Have interest in an actor, director, what the director has to say; look for in-jokes; attend film festivals. (Aficionado)
Knowledge of how films are created; study films; track what's being developed, release dates, critics reviews, and box office success; observe different directing styles; understand screenplays. (Technical Knowledge)

Do you ordinarily attend movies with a companion?
Avoid crowds; prefer to see a film without distraction from others. (Mostly/Prefer Alone)
Like to watch companions' reactions during the film; interact with friends during the movie with eye contact or comments; spend a few hours with friends before or after the movie; usually dinner or a cafe is involved. (Social)
Take dependents who chose the film, or there at the request of someone else; focused on the other(s) rather than on the film itself. (Facilitator)

FIGURE 5.5.

Define the non-obvious moviegoer audience segments with multiple-choice questions.

If a tally for a particular row grows too large, the spreadsheet is coded so that it will appear in a red font in the far right column. The recruiter will know that this candidate does not fit our needs, and can thank them and hang up. If you have the correct number of recruits for each audience segment, the right column numbers show up in black font, as shown in Figure 5.6. Additionally, a tally at the bottom helps the recruiter track how many of each audience segment has been recruited. Because the audience segments are composites of certain parts of the three aspects, Story, Craft, and Companionship, I include a sum at the bottom that indicates which of our four segments the candidate falls into.

Finally, I need to ensure that each candidate is easily going to keep up her end of the conversation for an hour. I want to hold a natural conversation with each candidate. So I add an "essay" question to each screener—something that requires an inventive answer and requires the candidate to talk for a few sentences. For example, I have used the question, "If you won $1,000, what would you do with it and why?" If the candidate has trouble answering, or uses a short retort, then I don't believe an hour-long interview will go smoothly (Figure 5.7). I ask the recruiters to dismiss this candidate. Another characteristic I ask the recruiters to gauge is how clearly the candidate speaks. In the past, I have had participants with such strong accents that I cannot understand them.

Ordinarily, I will hire a translator in advance and ask participants to speak in their native language. However, if I don't know whether a translator will be required, it is difficult to conduct the interview. If I'm interviewing internationally, I change this question to whether or not the person prefers a translator. I add a note to the recruiter asking if he agrees that the participant can speak clearly enough for the interview.

	Participant					
	101	102	103	104	105	106
(Check if we have required number for each segment. Check which						
Typical Moviegoer						
Craft Aficionado	3				3	
Story Fanatic		3	3			
Total Film Buff				3		3

FIGURE 5.6.

The "3" at bottom of each column identifies which of the four audience segments a candidate belongs to. Think of the "3" as a check mark. We want four from each audience segment.

If you won $1000, what would you do with it, and why? *(Judge whether they are articulate or not.)*
Articulate interlocutor
Short answer, not articulate
(Does the participant speak clearly?)
Speaks clearly
Hesitates, doesn't understand, garbled, cagey
The interview will be confidential & anonymous, but we will record it for transcription purposes. Is this okay with you?
Yes, recording is okay
No, please do not record me
Can you arrange a time in your schedule the week of September 13th for an hour phone interview?
Yes, arrange a time
No, cannot find time

FIGURE 5.7.

Extra questions ensure you will be able to hold an hour-long conversation with this candidate.

			Participant						
107	108	109	110	111	112	113	114	115	116

		3		3			3		3	4
3					3					4
		3						3		4
			3			3				4

Two logistical questions finish the screener. Because we want transcripts of what is said during the interview, we need to ask in advance if the candidate will permit recording the conversation. Then the recruiter asks if the candidate has an hour in her schedule during the week(s) of the interviews.

If the candidate meets our needs and has the time available, the recruiter can then set an appointment.

Coordinate Schedules

On a third tab in the same spreadsheet, I have prepared a daily schedule for the week(s) that my team and I will be conducting the research. Ahead of time, we each indicate our availability in the spreadsheet. Then the recruiter can set up appointments with the candidate on the phone by matching our availability with the candidate's availability. The recruiter simply adds the candidate's identification number in the date and time slot selected (as shown in Figure 5.8). The recruiter can also highlight what time zone the candidate is in, for our convenience when calling the participant.

	Time Available						
	Possibly Available						
	Not Available						
					Interviewers		
	PDT	MDT	CDT	EDT	Indi	Sarah	ID #
Tue	7:00	8:00	9:00	10:00			
14-Sep	8:00	9:00	10:00	11:00			
	9:00	10:00	11:00	12:00			
	10:00	11:00	12:00	1:00	Indi		103
	11:00	12:00	1:00	2:00			
	12:00	1:00	2:00	3:00		Sarah	114
	1:00	2:00	3:00	4:00			
	2:00	3:00	4:00	5:00			
	3:00	4:00	5:00	6:00	Indi	Sarah	115/104
	4:00	5:00	6:00	7:00			
	5:00	6:00	7:00	8:00			
	6:00	7:00	8:00	9:00			
	7:00	8:00	9:00	10:00			
	8:00	9:00	10:00	11:00			
	9:00	10:00	11:00	12:00			
Wed	7:00	8:00	9:00	10:00			
15-Sep	8:00	9:00	10:00	11:00			
	9:00	10:00	11:00	12:00			
	10:00	11:00	12:00	1:00			
	11:00	12:00	1:00	2:00	Indi		105
	12:00	1:00	2:00	3:00			
	1:00	2:00	3:00	4:00			
	2:00	3:00	4:00	5:00		Sarah	106
	3:00	4:00	5:00	6:00	Indi		108
	4:00	5:00	6:00	7:00			
	5:00	6:00	7:00	8:00			
	6:00	7:00	8:00	9:00			
	7:00	8:00	9:00	10:00			
	8:00	9:00	10:00	11:00			
	9:00	10:00	11:00	12:00			

FIGURE 5.8.
The recruiter assigns the ID number to an available interview slot, assigns the interviewer, and highlights the time zone of the participant.

Note that Figure 5.8 shows two appointments scheduled for the same time on Tuesday at 3PM PDT. There are two interviewers available at that time, so these twin appointments can be accommodated. In the spreadsheet I just ask the recruiter to put a slash mark between the ID numbers. This circumstance does not occur very often, so I feel that keeping the spreadsheet simple is the best solution.

World Time Clocks and International Dialing

When you deal with time zones that are more than four hours distant, it starts to get difficult keeping track of what time it is in both locations. You need to be sure you call the participant at the time they are expecting. While there are universal time clocks (e.g. UT, GMT, or Swatch's Internet Time), most people still use the time zones set up in the 19th century. There are several sites you can use to calculate what time it is in two locations. One of them is the World Clock Meeting Planner at **www.timeanddate.com/worldclock/meeting.html**

Another part of the puzzle is international dialing. Again, there are many sites that can help you with this. One that is related to the site above is **www.timeanddate.com/worldclock/dialing.html**

(Nitpick: Hopefully they'll modernize the interface some day.)

Also of note: I verbally request that the recruiters give me an hour between each interview. This gives me extra time if an interview goes longer than an hour or if we are late getting started. If you are driving between appointments, ask for the appropriate amount of travel time.

In Figure 5.9, you can see to the right of the ID# a column that lists the "Stipend Selected" by the participant, and the status of whether that stipend was sent. If we are giving out gift cards at the end of each interview, we give the participant a choice of several merchants. For the moviegoers, of course, we offered movie gift-certificates, in this case from Fandango. We also offered gift cards from Home Depot, Starbucks, and Target. After the interviews, someone else on the team looked at this spreadsheet, referred to the addresses recorded on the Recruits tab (see Figure 5.10), handed off, emailed or mailed the gift card, and marked the status on the Schedule tab.

	Time Available								
	Possibly Available								
	Not Available								

	PDT	MDT	CDT	EDT	Interviewers Indi	Sarah	ID #	Stipend Selected	Stipend Sent
Tue	7:00	8:00	9:00	10:00					
14-Sep	8:00	9:00	10:00	11:00					
	9:00	10:00	11:00	12:00					
	10:00	11:00	12:00	1:00	Indi		103	Target	sent
	11:00	12:00	1:00	2:00					
	12:00	1:00	2:00	3:00		Sarah	114	Target	sent
	1:00	2:00	3:00	4:00					
	2:00	3:00	4:00	5:00					
	3:00	4:00	5:00	6:00	Indi	Sarah	115/104	Starbucks/Fandango	sent/sent
	4:00	5:00	6:00	7:00					
	5:00	6:00	7:00	8:00					
	6:00	7:00	8:00	9:00					
	7:00	8:00	9:00	10:00					
	8:00	9:00	10:00	11:00					
	9:00	10:00	11:00	12:00					
Wed	7:00	8:00	9:00	10:00					
15-Sep	8:00	9:00	10:00	11:00					
	9:00	10:00	11:00	12:00					
	10:00	11:00	12:00	1:00					
	11:00	12:00	1:00	2:00	Indi		105	Fandango	sent
	12:00	1:00	2:00	3:00					
	1:00	2:00	3:00	4:00					
	2:00	3:00	4:00	5:00		Sarah	106	Fandango	sent
	3:00	4:00	5:00	6:00	Indi		108	Home Depot	sent
	4:00	5:00	6:00	7:00					
	5:00	6:00	7:00	8:00					
	6:00	7:00	8:00	9:00					
	7:00	8:00	9:00	10:00					
	8:00	9:00	10:00	11:00					
	9:00	10:00	11:00	12:00					

FIGURE 5.9.
Tracking the stipend requested by the participant, and whether or not it has been sent.

The Recruits tab is where the recruiter lists the participant information. It is the *only place* the participant's name and contact information appears (Figure 5.10). Everywhere else, including the transcripts, the participant is referred to by ID number. This practice helps ensure confidentiality. However because ID numbers are so unmemorable, my team and I usually make up nicknames for each participant so we can refer to them more easily. Under no circumstances do you want to propagate their true names.

ID #	Name	Audience Segment	Notes
	**Selected Participants**		
101	Kat	Craft Aficionado	Lives alone near downtown art theater
102	Jerry	Story Fanatic	New college graduate, lives in city
103	Marie	Story Fanatic	Urban, mom of toddler, likes to watch DVDs at home
104	Larry	Total Film Buff	Former film student, wants to write movie
105	James	Craft Aficionado	Goes to movies with wife every Friday night
106	Joseph	Total Film Buff	Sees all movies on opening night
107	Jeffrey	Craft Aficionado	Lives alone, no TV, works for special effects company
108	Daniel	Story Fanatic	Single dad, works a lot
109	Gregory	Typical Moviegoer	Probably pretty average movie-goer
110	Juan	Total Film Buff	Actually making his own copy of original Star Wars
111	Jennie	Typical Moviegoer	Newlywed, go to movies every Sunday with mother-in-law
112	David	Craft Aficionado	Lives alone, sees movies with friends
113	Rich	Total Film Buff	Has home theater, organizes outings to movies w/ friends
114	Carol	Typical Moviegoer	Mother of three, husband orders movies on Netflix
115	Lowell	Story Fanatic	Creates CDs based on soundtracks his 2-year-old likes
116	Jeanne	Typical Moviegoer	Retired, watches movie channel with retired husband

FIGURE 5.10.
The Recruits tab of the spreadsheet is the only place where confidential participant information, including their names, resides. (Note: I changed the data in this image to protect the participants' identities.)

On the Schedule tab of the spreadsheet, there are columns that represent each interviewer. If you would also like to schedule translators, add columns for each of them and have them mark their availability. This method ensures a translator will be available for each interview. If you would like to invite stakeholders or possibly the participant's sales representative to be present for the interviews,[1] show them the schedule and let them select the interviews they can attend.

Recruit Participants

As I mentioned earlier, there are a few methods of recruiting. You can use a pop-up on a web site to attract participants. You can get a database of phone numbers and call each candidate. You can ask sales reps to find people. You can pick on friends of friends, using your personal network. There are various reasons to use each method, as detailed in Chapter Six of *Observing the User Experience: A Practitioner's Guide to User Research*

---

[1] Read more about "ghosting" in an interview in the Chapter 7 section, "Ghosts and Ownership."

by Mike Kuniavsky. Mainly, I encourage you to hire a recruiter when you want to recruit people you don't already know. Even if you have a huge database of names, but you don't have a dedicated internal person who can phone them all and ask them screening questions, hire a recruiter. A dose of reality: Recruiting could take someone eight hours a day for eight days for a pool of 16 people.

Recruiting Gave Me Gray Hair

"This takes ABSOLUTELY ALL MY TIME. It's more than phone calls. It's massaging the reps, establishing relationships, chit chat, scheduling. I have gray hair now!"

—Andrea Villa, Senior Project Analyst, Qualcomm

It's not easy. Recruiting agencies do require daily management by someone on your team, but it's closer to one hour a day for eight days. Managing data from a pop-up and contacting candidates for verification of qualifications and setting up appointments will take a person three hours a day for eight days.

That said, it is difficult to find a good recruiting agency. No matter how well the project manager on the agency's side understands that you want to hold in-depth conversations, this understanding sometimes doesn't trickle down to the people making the phone calls. During every project I find at least one recruit who doesn't quite fit the criteria. I point this out to the agency, and they replace that recruit with someone better. A good agency won't charge for replacing a poor fit. Sometimes, though, you will not know the recruit is a bad fit until you're 10 minutes into the interview. When you decide to cut the interview short and ask for a replacement recruit, know that you will be charged[2] for the replacement, as it was impossible to tell the original person was a bad fit just from the screener.

---

[2] I hesitate to write down the average cost for recruiting because you may be reading this many years in the future. But for ballpark purposes, in 2007 it costs between $100 and $150 per participant.

You may even feel obliged to give the participant the stipend, even though you aren't going to use their transcript.

If you use the pop-up method to recruit candidates, lots of your work is done for you. However the basic premise that you will need a site or two (for audience variety) to base your pop-up on may not be easy to achieve. Perhaps you don't have—or shouldn't use—your own site to find recruits. In that case you will have to negotiate to use another site or two as a base. Moreover, there is the problem that candidates will self-select with extraordinary fervor if your stipend is attractive—that is, people will gladly lie about their answers on the screener to get on your list. You may have to create a secondary screener to ask these candidates over the phone, crafted to tease out the truth. It's difficult to do. In the case where I wanted people who were two to five years out of college, I got at least four participants who had just graduated. The truth comes out during the interviews when participants are less guarded about what they say.

The Human Skull Collector, and Other Stories

Once I was interviewing people who ran small businesses. The recruiting agency was tasked with finding as much variety as possible. The stipend was high, since we needed to distract these business owners from their businesses for a few hours. My conversations with people were pretty interesting. Then this big guy with glasses and a long ponytail walked in. I opened up the conversation, asking him about his business. "I'm a human skull collector," he said. "Excuse me?" I asked, figuring that my hearing was off after so many interviews that day. "Human skulls. Genuine human skulls. I collect them." Good lord! Tentatively I asked, "And how have you made this into a business?" He quirked a smile and said, "Oh it's not a business. It's a collection." I went on to ask him about what business he was in, and it turned out he wasn't in any business at all. So with much relief, I thanked him, gave him his stipend, and sent him on his way. I definitely didn't want to ask him how he collected the skulls—eBay or something. I hope.

In another instance, I was showing some prototypes to people for some evaluative research. Again, we were recruiting small business owners and the stipend was high. One fellow came in and I showed him the screen and asked him a few questions, telling him to point where he was looking so I could see what he was talking about. Right off the bat, he wouldn't point, and he couldn't say what he'd use the application for. I tried a different screen, and again, no pointing. I began to suspect the fellow couldn't make out what was on the screen. I tried enlarging the display and asked again. The fellow still couldn't tell me what he'd do with such a screen. I pointed to a heading and asked him what he thought that area was for. He shrugged. I suddenly realized that the poor guy probably couldn't read very well and was too embarrassed to admit it. "Ohmigod," I thought to myself, "Do I have to put 'literate' on these screeners as well?" Then again, how would you know, if you're recruiting someone by phone, that the candidate couldn't read? So, I thanked the guy, gave him his stipend, and showed him out.

In any case, the pop-up can act as your screener. Leave off the extra questions shown in Figure 5.7. If a potential candidate's answers indicate he is not a good fit, show a screen thanking him for his time and telling him you already have enough people matching his criteria. If he is a good fit, get his contact information and call him to validate his answers, ask him the essay question, check his communication skills, and schedule the interview. Do your best to let them know what to expect. In many cultures, candidates want an official letter[1] describing your organization and the extent of the interview. Your goal is to make people feel comfortable with the information they will give you during the upcoming interview.

If you have a set of audience segments that are rare or difficult to find, and you have internal customer or sales lists that fit the requirements, you may want to recruit from these lists. Ordinarily, a recruiting agency has developed their own database of names, and they recruit from this database when you hire them. Therefore, if you're going to use your internal lists, you'll either have to find someone internally to do the recruiting, or you'll want to negotiate with the recruiting agency to use your lists for a reduction in their overall fee. When using a database of names, the "hit rate" (number of candidates who fit your criteria and want to participate in an interview) can be as low as 2%. Therefore, if you are looking for 16 recruits, you might need to start with a list of 800 names. Many large companies keep track of how many times each customer is contacted each year, and mete out lists of people you can phone in sets of 100. Be prepared to work with the gatekeepers of that database to get a better pre-filtered list, so they don't have to mark a big swath of 800 people as "already contacted" for the year.

While recruiting is going on, you can use the time to explore the scope of what you want to cover in the interviews and craft some guiding prompts. The next chapter will guide you through this process.

---

[1]   See "Introduction Letter to the Interview Participant" in the Resources section of the book site. www.rosenfeldmedia.com/books/mental-models/content/resources

CHAPTER 6

Set Scope for the Interviews

Set Research Goals   94

List Interview Prompts   96

You might have a week, you might have an hour. What will you talk about?

A technique that I frequently use is a short, non-directed interview. I have included this technique in the next two chapters of this book. This first chapter outlines how to scope your research and get into the right frame of mind. The second chapter shows you how to conduct the interview itself.

You will want to ensure that you cover topics that are of concern to your team. You also will want to keep the conversation during your interviews from straying into areas that aren't of particular interest or relevance. Use your time wisely.

Set Research Goals

Try to understand what the organization is trying to get out of the solution you'll be designing. You will find yourself scoping and re-scoping your research so you know you're really pinpointing what stakeholders are interested in. You need to know what you want to study before you go out and start listening to customers. Start by reviewing the goals of the solution you are developing:

- What is its intended purpose?
- What do the leaders of your organization want out of the product?
- What advantages and difficulties does your team see?

The project leader should also interview stakeholders about the organization's mission, long-term strategies, objectives, predictions, perceived challenges, places where things didn't go well in the past, strengths and weaknesses of the internal structure, and who the decision-makers are. Record the interviews. Look for patterns in what various business stakeholders say, and reflect those patterns back at them to validate that you understand. Sometimes the patterns you find will contradict each other. Be sure to follow up on incongruities and ask the stakeholders for a consistent direction—or at least make them aware there are incongruities. Remember these patterns will define the scope of your project.

After you have interviewed all the stakeholders of the project, group their goals together into overarching themes. Then review the research scope that you set when doing the audience segmentation. Cut any stakeholder themes that fall outside the scope you defined. If a theme seems too important to cut, examine whether it is worth extending the scope, and possibly adding a different audience segment (Figure 6.1).

GOAL	PRIORITIZED KEY BUSINESS OBJECTIVES
Simplify Access to Information	1. **Simplify Web Navigation** 2. **Have One Site** 3. **Make Things Consistent** 4. **Base Navigation on Audience Needs** 5. **Improve Look & Feel** 6. Resolve Access to Large Information Stores
Support Decision Makers with Better Information	1. **Support International Customers & Sales** 2. **Understand Users' Needs** 3. **Improve Depth of Technical Material** 4. **Tag/Purge Old Information** 5. **Support All Stages of the Buying Process** 6. Expose More Good Internal Material 7. Give In-Depth Explanations/ Transcripts for Presentations 8. Expose Governance Material
Save Company Resources	1. **Allow Customers to Answer Questions Themselves** 2. Allow Customers to Track Inventory, Backlog & Orders 3. Streamline Web Content Creation & Maintenance 4. Eliminate Printed Sales & Marketing Materials 5. Allow Customers to See Their Satisfaction Data 6. Attract Qualified Staff

FIGURE 6.1.
Stakeholder goals for a multi-national organization, where goals applicable to the customer web site redesign are highlighted with bold font. Those items not in bold font are ignored for the scope of this project.

Next, you will want to look at user research reports from prior efforts, from within your own department and from other divisions as well. These reports will give you a greater depth of understanding about where the organization has explored the whole user experience in the past. Look for usability reports, focus group findings, customer satisfaction surveys, problem reports, web traffic analysis, and so on. Any prior knowledge, even if it seems a little tangential, will give you a broader awareness.

Mental models can depict different levels of the same thing, and these levels are slippery to define. Don't try capturing two levels in one model. Enunciate which level you're after up front. Often I will start with one scope, say "remodeling," and revise it as I start to interview people to something more specific, like "touch points remodelers have with people, locations, and information." I am sure up front, though, that I am not exploring the lower level: "steps to remodel a kitchen." (Also see "What Level of Granularity Should I Go For?" in Chapter 7.)

After I have conducted an interview or two, sometimes I feel the need to adjust the scope of the conversation. Maybe I've heard about a topic I wasn't expecting, or maybe I realized that one of the topics I wanted to cover makes no sense in context. Since this is qualitative research, don't worry about changing the conversation of subsequent interviews a little. It won't affect the outcome of your research.

List Interview Prompts

At this point, remind yourself and your team that mental model research is qualitative, not quantitative. The questions you will create are not asked the same way to each person. If you can't compare answers for each question from each person, you can't derive any statistical significance from your data. Rather, each conversation is unique and follows its own path. In this way, you can truly hear what each person says rather than coax their conversation to fit a pattern you, consciously or subconsciously, have in your head. What you are doing explores systematically "the why and how of decision making."[1] When performing qualitative research, it is important that you approach your participants with as little personal baggage as possible.

Leave Your Own Beliefs Behind

In Dan Saffer's* book, *Designing for Interaction: Creating Smart Applications and Clever Devices*, he interviews well-known researcher and speaker Brenda Laurel, PhD.** He asked her what designers should look for when doing research. Her answer emphasizes the importance of shedding assumptions and precepts before asking research questions.

"The first step is to deliberately identify one's own biases and beliefs about the subject of study and to 'hang them at the door' so as to avoid self-fulfilling prophecies. One must then frame the research question and carefully identify the audiences, contexts, and research methods that are most likely to yield actionable results. Those last two words are the most important: actionable results. Often, the success of a research program hangs upon how the question is framed: 'why don't girls play computer games?' vs. 'how does play vary by gender?'"

---

*   Read more of Dan's writing at www.odannyboy.com

** Read more of Brenda's writing at www.tauzero.com/Brenda_Laurel

Non-leading interviews allow you to capture what a person is thinking in their terms, with their structure and vocabulary intact. When you craft the list of topics you will cover during your conversation, you should be as nonspecific as possible. I deliberately write prompts instead of interview questions. I write them in a terse phrase format without a question mark at the end. I do this purposely so that I will not be inclined, during the pressure of a real interview, to retreat to rote—that is, to read vocabulary

---

[1]   *Wikipedia's* definition for "qualitative research," whereas quantitative research is defined as "the what, where, and when."

that I may not have heard from the interviewee, or to recite pre-specified questions. These prompts force me to keep up with the conversation—to stay engaged and actively thinking about what I want to explore further. (For an example of prompts, see Figure 6.6. In the next chapter, I will help you practice non-leading interview techniques.)

Another advantage of having a list of prompts during an interview is that they are easier to parse quickly. Your eyes can sweep down the list. If you are at the end of one topic of conversation, and you need to direct the next step, a certain noun or phrase will automatically jump out at you from the prompts. During the hectic pace of an interview, any sort of mental crutch is welcome.

Assemble Topics

When you have your list of product themes from stakeholders, brainstorm keywords that have to do with each theme. Use phrases or nouns. In Figure 6.2, I show examples under four moviegoer themes.

Choose the Movie
- find out
- reviews/opinions
- plot
- director/actors
- read the book

Choose the Theater
- playing times
- locations
- parking/transportation

At the Theater
- tickets
- concessions
- seat
- experience while watching
- companions
- habits

After the Movie
- reviews/opinions
- companions
- habits
- food
- unique relationship to movie

FIGURE 6.2.
Associate keywords and phrases with each theme.

Next, list your audience segments. Ask yourself what keywords are specific to each group, as in Figure 6.3.

Typical Moviegoer
- someone else chooses
- react out loud
- discuss whether I like it or not

Story Fanatic
- opening night
- craft of film making
- genre
- write review
- follow film industry

Craft Aficionado
- opening night
- discuss craft of film
- leave film early
- variety of films
- watch others for their reactions
- good audience makes the film

Total Film Buff
- observe craft of film making
- never leave film early
- variety of films
- see what's playing when I want to go
- refuse to stand in line
- follow film industry
- prefer theaters with character
- attend films alone

FIGURE 6.3.
List keywords and phrases with each audience segment.

Look at all your phrases, both for the themes and the audience segments, and determine whether certain sets apply to different types of people you will interview, or if you could ask about most of the topics to each of the interviewees. Organize the prompts accordingly. I often write the prompts into a deliverable that also describes the types of people I will be interviewing. I send this deliverable to the stakeholders to make sure that the scope I have defined with the topics matches what they expect from the research. For my own use, I jot the prompts in little notes to have available during the interviews, as you will see later in this chapter.

Prepare Reminders

Because I do a lot of my interviews by phone, I like to jot down the number to dial, any pass codes I need, and the instructions for turning on and off the recording. I write a reminder to myself to check if anyone else has dialed into the line to listen to the conversation (Figure 6.4). Once

I get the participant on the line, I like to give them an explanation of what kind of research we're doing and why we're having a conversation. I reassure them that the interview is confidential and that no one else will be reading the transcript but my team. I make sure that they're comfortable with an hour long conversation. Moreover, I remind them they will be getting paid for their time. Then I ask permission to record the interview. If I receive permission, I turn on the recording and ask for permission again, so that I have their response in the audio file.

Greeting

Dial the conference facility toll free number_____
- Make sure the recording is off (*2)
- Find out if any ghost participants are on the line
- Dial out (*1) and then the participants phone number
- Dial back in (*2) to the main conference line

- I am with_____, a research firm.
- Our goal is to learn how you _____.
- This is a confidential conversation.
- (If ghosts) There is a person listening in for research purposes.
- The interview will last one hour.
- We will send you a $____gift card after the interview.
- Are you in a comfortable place to speak for an hour?
- Do we have your permission to record this conversation for transcription purposes?
(If so, turn on recording (*2) and ask for the record.)

FIGURE 6.4.
List reminders for the initial greeting for interviews, as well as dialing information.

Recording a phone conversation is simple if you use a conference call service. Usually there is a built-in recording capability that you switch on and off with commands on your telephone's keypad. These recordings are usually high quality. Sometimes there is a time lag before you can access the file containing the recording, but usually the files are immediately

downloadable from the service provider's site.[2] If you are dialing someone
directly instead of using a conference service, you can connect a recording
device to your phone. I have seen such devices at Radio Shack[3] for a
reasonable price, and doubtless you can order them from other vendors
online. You use them with recording software (try one of the free MP3
recording suites) that will save the file on your hard drive. For convenience,
I recommend recording to a digital file rather than to some other media,
like audio tape. A file is a lot easier to store and distribute. If needed,
it's relatively easy to convert the file format to a version your transcriber
can listen to. If you are interviewing the participant in person, bring the
recording device with you, but test it first to be sure that the volume is set
correctly. Nothing is worse than conducting a really good interview, then
finding out later that you had the recording device on mute or turned down
too low.

I begin the interview with a softball question—a question that is simple to
answer and puts the participant at ease. I jot down ideas as reminders for
myself during the conversation (as shown in Figure 6.5).

Softball

- (If single) Please tell me a little about what you've been doing with
  your dating life this month.
- (If taken) So you're in a relationship. How did you two meet?
- (If taken) Please tell me a little about how you've been involved this
  month with your friends who are dating.

FIGURE 6.5.
Softball questions get the conversation rolling during interviews about
dating and matchmaking for Engage.com.

---

[2] For the convenience of immediately downloadable audio files, I use the AccuConference service.
www.accuconference.com There are many other services similar to this provider. Be sure to
choose one that allows you to dial out to bring the interviewee onto the line with the rest of
your team.

[3] www.radioshack.com

After the participant has started the conversation, I just follow it from there. I will refer to my prompts list from time to time, but it is not where I focus (Figure 6.6). (If you really want to know, I look out the window so I can concentrate on what the person is saying.) Occasionally I will jot a word down as a reminder to follow up on a certain branch of the conversation. When a topic peters out, I will just move on.

I recommend going over this list of prompts with the team who will do the interviews to make sure everyone understands the intent behind each topic. Differences between things like "debrief with friends" (assess the date) and "discussion with friends" (talk about relationships) might not be obvious based on the way the prompts are worded. Review the list with the team and edit as necessary until each line is crystal clear, yet still succinct.

At the end of the conversation, I write a reminder to myself in big bold letters to turn off the recording, because I don't want to capture the contact and stipend information in the file (Figure 6.7). I conclude with thanks and the details of how the participant will receive the stipend. I often double-check the participant's contact information at this point and tell them whom to contact if they don't receive their stipend after two weeks.

```
Prompts
----------
If Single
```

- Advertise/Hide the fact you are available
- Initial attraction (criteria), beyond looks, time constraints, humor
- See no one compatible
- Let someone know you're interested/not
- Find out more about a person before a date
- Prepare for a date (mentally)
- Go on a date (discussion) (internal dialog) (friends involved) (let know interested/not)
- Etiquette/manners
- Debrief with friends after a date (male/female)
- Discussion with friends/others, emotional expressions
- Advice, advise, reaction, change
- Put in a word on friend's behalf
- Point out someone to a friend
- Who to hook up together (why)
- Introduce someone
- Change over the years
- Stop dating for a while

```
If Taken
```

- Picking someone out for a friend (criteria)
- Find out more about a person for your friend
- Putting in a good word on friend's behalf
- Debrief with friends after their date (male/female)
- Discussion with friends/others
- Advice, advise, reaction, change
- Other taken friends of yours (bring them in)
- Point someone out to a friend
- Who to hook up together
- Introduce someone
- Change over the years

FIGURE 6.6.
Jot down prompts for reference during interviews.

> Conclusion
> ----------
> - TURN OFF recording. (*2)
> - $___gift card
> - Check address
> - If you do not receive this within two weeks, please email the project manager at _____@_____

FIGURE 6.7.
Remind yourself to shut off the recording before discussing stipend and contact information.

If having these little notes around while conducting the interview might distract you, then don't do it. If the notes provide too much of a crutch by indulging your tendency to recite what you've written, then don't do it. The idea is to help you concentrate on the conversation.

Easily Flustered?

It's easy to let your nerves get the better of you when setting up for an interview. I confess to several mistakes, like dialing the interviewee but asking for the wrong person—I was looking at the name on the next line of the spreadsheet—or asking a trucking dispatcher to tell me how the day starts as a mechanic in the garage. I have also made the costly mistake of dialing out to a participant (hitting *1 on the telephone keypad, as shown in Figure 6.4), but forgetting to join back into the main conference line (by hitting *2). I conducted the entire interview while ignoring the incoming calls on my mobile phone from teammates desperate to know why I had left the conference line dead. It was a good interview, and none of it was recorded.

Indeed, the conversation is everything. The next chapter should address your worries about directing a non-leading interview.

CHAPTER 7

Interview Participants

Chat by Telephone or Face-to-Face   106
Do Not Lead   107
Plan Your International Interviews   125

Many people have written[1] about how to conduct a non-leading interview. The skill is required of so many disciplines: ethnography, market research, management. You need to think hard while listening to the interviewee, recognizing when the conversation is non-productive, and then nudging it back on track. You need good social skills, so you can make the interviewee feel comfortable speaking to you. You need to be aware when you hit sensitive ground in the conversation. Above all, you need to be able to get the information you need out of the discussion.

Chat by Telephone or Face-to-Face

First of all, I'm going to tackle the question of whether to interview someone in-person or remotely. I take the following approach: If there is a *unique* artifact a person uses, a context in which she acts, or group feedback she gets that *influences* the kinds of things she does, then I want to see her in person. Otherwise, I conduct interviews by telephone. Because I am not conducting evaluative research, I don't need to see how people behave when interacting with their tools. That kind of detail is too specific for a mental model and would get grouped in with a higher-level concept. I just need to know the general steps people take. For that, a verbal description is fine.

There are many advantages to site visits and direct observation, and there are just as many advantages to remote research. Most obviously, it costs less because no travel is involved. You also reap savings by bracketing your research into hour-long time periods, rather than day-long site visits. In this way, you gather just as much data as you need, then move on. If the process you are exploring for the mental model spans months, the participant can summarize the parts that are important to her in a matter of minutes. Additionally, getting a global understanding of potential customers for a product is important. Conducting interviews by telephone allows you to interview in every location you can conceive of people using

---

[1] Chapter 10, "Conducting the Site Visit—Honing Your Interviewing Skills," by JoAnn Hackos and Janice Redish in *User and Task Analysis for Interaction Design*; Chapter 6, "Universal Tools—Recruiting and Interviewing," in Mike Kuniavsky's *Observing the User Experience*; and Chapter 4, "Contextual Inquiry in Practice," in Hugh Beyer and Karen Holtzblatt's *Contextual Design*.

your product. This advantage alone is of immense power for
many products.

Building a mental model is generative research. It is an exploration of
information, as the participant knows it. You are allowed to ask participants
to tell you about it rather than strictly observe them.

In-Person Interviews of Media Buyers

In 1998 I conducted a set of interviews for a dot-com startup interested in
providing an online solution for media buyers. Media buyers are professionals who determine where to place what kinds of advertising for a particular
client/event. For example, the media buyer for the San Francisco Giants
might place season-oriented ads on bus shelters and specific game-related
ads on the radio the week before the game. Each media buyer that I was to
interview had a different way of tracking ads and data and making decisions
based on that data. Some used complex Excel spreadsheets. Others used
a set of folders with ad copy and printouts. Each of them approached the
problem with a slightly different set of steps. I needed to see my interviewees
in person in order to know to ask about the unique tasks they were doing.

Do Not Lead

If you're not going to lead, you still need to stay on your toes. I am not a
dancer, but perhaps this rule applies to the ballroom floor as well. You need
to stay focused on where your conversation partner is moving.

Think of your interview as dinner party conversation. You have been
seated next to someone you do not know. You don't have a list of questions
scribbled on the napkin on your lap to peek at, so you start out with some
typical topics. "How do you know our hostess?" "What do you do for a
living?" "What kinds of pastimes interest you?" "What is your favorite
movie, book, music, etc.?" These are the socially acceptable "prompts" that

we've all learned from experiences talking with strangers at parties. You don't have any expectations of the person's responses. You ask for more detail about things the person says as the conversation progresses. You follow the tangents when the person switches topics. You act interested. If the person starts talking about a topic you don't want to pursue, you change the subject. Your goal is to get to know that person, pass the time at the party enjoyably, and possibly make a new friend.

The fact that you have no expectations about the person's responses is important to how you should approach a non-directed interview For illustration, I want to talk about two examples of the opposite technique—when the person wants the conversation to fit into some structure or belief they have already formed. You probably have experienced the interlocutor who forces the conversation back to his own points again and again. This person is insisting that you shift your worldview to match his own. As you know, these are uncomfortable conversations. (Don't confuse this example with a person who plays devil's advocate so that the two of you can both explore different perspectives together and enjoy learning new points of view.) A second example is a TV or radio personality who conducts interviews. This person has the explicit goal of entertaining his audience by drawing out the interviewee, trying to get them to reveal something new, talk about something controversial, or tell a titillating story.[2] In both of the above examples, the interviewer directs the conversation to the topics he wants to talk about and asks questions about things that have not been mentioned yet by the interviewee. You want to do the opposite.

Of course, if the conversations starts heading in a direction that is out of scope for your research, you will want to redirect it. Keep your ears open for this. Above all, though, you want to feel confident and relaxed. Don't expect too much of yourself. Running a non-directed interview is difficult at first, so you have permission to make a few false starts in each conversation.

---

[2]   Terry Gross's interviews on National Public Radio's "Fresh Air from WHYY" (www.whyy.org/freshair/) are examples of interviewing with the goal of entertainment in mind.

Six Rules for Mental Model Interviews

There are six rules for mental model interviews that I've collected over the years that help when learning the art of non-leading conversations. I introduce them here and explain each in detail.

1. Behaviors and philosophies, not product preferences
2. Open questions only
3. No words of your own
4. Follow the conversation
5. Not about tools
6. Immediate experience

1. Behaviors and Philosophies, not Product Preferences

From the beginning, the main mistake I wanted to avoid was following what most software companies' marketing departments were doing in the 1980s and early 1990s. The company would have a product under development, and marketing would be partially in charge of figuring out the next feature set. To do this, they would ask current customers what they liked and hated about the product they were using. They would ask what new functionality the customer would like to see in the product. Then they would report their findings to the engineering department. It was a very narcissistic way to develop a product. So, I resolved not to ask customers about the product at all. I wanted to find out, instead, what customers were trying to achieve. I even talked to people who weren't customers, to see if people who weren't early adopters had different goals in mind. So my first rule about interviewing was to concentrate on what the person is trying to do and not on the product itself. Avoid the words "love," "hate," and "wish." This rule moves you from the old type of evaluative research done by marketing groups to the generative research category.

I used to refer to this rule as "Do, Not Feel," but people confused feelings with philosophies in their interviews, thereby missing out on the latter. By "feel" I meant "love it," "hate it," or "wish it did this." I changed the rule to "Tasks and Philosophies, not Product Preferences" to be clearer...not to mention the fact that people could misinterpret the previous wording as a demand: "Do not feel." Or something that Yoda would say, "Do not; feel."

2. Open Questions Only

Remember your junior high school journalism class? The six journalist questions were drilled into your head: who, what, when, where, why, and how. There is a good reason for this: These are open questions. A good reporter won't let her own opinion prejudice a report she is writing. By asking "Were you at the Ivory Tower Nightclub last Saturday night?" she indicates she believes you were there, and she is missing the opportunity to find out where you were if you answer, "No." By asking "What did you do last Saturday night?" she will get a more frank answer, and not let her interviewee know what she has pre-decided in her mind. Begin your questions with these six words and you can't go wrong. Begin with "Did," "Have," "Are," "Were," or "Will" and you are leading the participant down the path of your assumptions.

Ask Open-Ended Questions

In the book *Difficult Conversations* by Douglas Stone, Bruce Patton, and Sheila Heen of the Harvard Negotiation Project, there is a chapter called "Listen from the Inside Out." One of the paragraphs on page 174 of the Penguin edition makes this point:

"Open-ended questions are questions that give the other person broad latitude in how to answer. They elicit more information than yes/no questions or offering menus, such as, 'Were you trying to do A or B?' Instead ask 'What were you trying to do?' This way you don't bias the answer or distract the other person's thinking by the need to process your ideas. It lets them direct their response towards what is important to them. Typical open-ended questions are variations on 'Tell me more' and 'Help me understand better...'"

3. No Words of Your Own

This is the rule that raises the most eyebrows. How on earth can you interview someone if you can't use any vocabulary that she hasn't already uttered? Well, it's possible. Of course I don't mean common words that we speak all the time. When I say "no words of your own," I mean vocabulary that is specific to a topic or industry or circumstances of your product. Don't ask the media buyer "How do you determine your strategic market?" if he hasn't mentioned "strategic market" yet. Maybe he calls it something else, or maybe he thinks of it as two distinct sub-components that he refers to in a different way. If you start throwing "strategic market" around in the conversation, he is liable to just shrug and go with what you understand, rather then explain his own perspective.

There is an oh-so-human tendency for participants to agree with you even though the structure you're voicing isn't necessarily their structure. This behavior is probably an attempt make sure the connection happens on the part of the participant. The person nods her head and thinks, "Yeah, yeah, that's close enough, so yeah, I agree." Avoiding vocabulary that hasn't yet been introduced by the participant will solve this tendency.

4. Follow the Conversation

This is the "dinner-party conversation" rule. Don't conduct the interview as a series of question-and-answer, call-and-response snippets. Don't act like an automaton reciting survey questions. Allow your participant to direct the conversation most of the time. If you must segue into a new subject, do it with a reference to what has been said so far. Have a conversation. Make sure your participant does most of the talking.

Since you are holding a conversation with the participant, you can adjust the flow of the conversation based on what the participant says. There is no need to cover your topics in any specific order, so don't worry if the participant brings up something that you had planned to talk about later. Dive into it when the topic is brought up. It results in unbiased data when you let the participant lead the conversation this way.

5. Not About Tools

Don't get into particulars about how the person operates each tool they use. This tendency is hard to overcome for those of us who have conducted a lot of usability research. Ahem. Find out what they are using the tools to accomplish—what's going on in their minds. What were they thinking when they were walking down the hallway to use this tool? (That's the "Hallway Test," found in Chapter 8.) Rather than dwelling on how a person looks up a phone number to make a call, explore what the call is meant to achieve. "Check in with a subordinate for status on a project?" "Place a re-order for a particular component?" Your conversation is not about the tool; it's about the course someone is following to get something done.

6. Immediate Experience

In Chapter 5, when I talked about recruiting, I mentioned the need to ask for people who have recently done the thing you are researching. During the conversation, you need to keep this in mind. Ask about "the last time you did this." It's easy to follow the conversation into the deep past, or into an area your participant does not have much experience with. Be aware of this when it happens and wander back to areas where the participant's intentions are strong and easily explained. I should mention here that ethnographers find that details are often misreported. They believe that site visits and observation are the best way to avoid getting fiction mixed in with your facts. However, at the high level that you're interviewing, there are few of these kinds of details in the first place. You're not interested in whether the person smiles before or after saying "thank you" to the cashier in the grocery store. You're interested in whether checking out is actually something that registers in the person's mind, or if "getting a good deal on each item" is something he concentrates on more.

What Level of Granularity Should I Go For?

Take the Goldilocks approach: not too much and not too little. You want the "just-right" level of granularity. I realize that is a vague description, so here is an example. If you are interviewing someone about what they do when they get up in the morning, that person might "mentally review the points I want to make in the meeting at 9AM" and "brush my teeth" and "get dressed." That person might also mention "ensure Johnny gets to his 8:30AM dentist appointment." You will want to focus on the major goals, not the minor ones. Getting dressed does not count as a major accomplishment for most people. Then again, for a particularly fashion-aware person, choosing just the right ensemble and checking the look in the mirror would count as a valid set of tasks. In either case, "Buckle my shoes" does not count.

Usually the focus is on the more important goals: the 9AM meeting or the 8:30AM dentist appointment. For the latter, if you dive in, you might discover goals like "Map out route to dentist with least traffic" or "Remember to bring signed release form." But don't ever go to the lowest level of granularity. You don't need to explore "Find pen to sign release form" or "Type URL for the traffic site." That kind of detail is not needed.

The mental model will be used to compare existing design features to what the user is doing, in order to decide if you are supporting people correctly. It will also be used to synthesize new design features. So the level of detail in the mental model doesn't need to touch on details, but it does need to encompass the spectrum of people's behavior and go into depth about the reasoning behind the actions.

Bolster Your Confidence

These six rules are what keep me confident and organized going into an interview. But rules are not everything. Experience is most of the equation. Before conducting this kind of interview for the first time, test it with a friend—say your running partner. Next time you're both out running together ask her what she uses her sports watch for. See if you can ask open questions and avoid introducing ideas or vocabulary that she hasn't mentioned yet. Try to get at the root reasons why she does things like "time my run." Let her lead the conversation. If the conversation veers into how little control she has over the time meetings take at work, gently shift it back to the scope of the sports watch. Press her for details about how she learned that "slowing the tempo of a run burns more fat" and why it is important to her. This trial interview will help you get the feel for the kind of conversation you want to have.[3]

I want to emphasize that you should *follow each topic through to its task-related conclusion.* For example, in one project we ended up with the phrases "manage employees" and "hold meetings." These are not tasks. Why are you holding meetings? What are you trying to get done? What, exactly, does it mean to manage an employee? "Set assignments" and "Encourage skill growth" are more like it. Just because you have assumptions about what it means, don't move on when you hear "manage employees." If it's within the project's scope, pursue it.

Be Considerate of Your Participant

Another thing to mention is sensitivity. You will be interviewing strangers, and sometimes the topic of conversation will touch on something they hesitate to discuss. Please be attuned to this and offer reassurance that you're not interested in their trade secrets, or simply apologize and move on to a different subject.

Sometimes a participant will wonder why this interview is so different than what he was expecting. A couple of times, participants have stopped me and asked, "Why are we talking about this? Don't you want to know how I use the ___?" I stop and explain to them the nature of

---

3 By the way, I'd love to help someone design a better trail running watch. There are some behaviors that are, as of yet, unsupported, like, "Decide Whether to Turn Off Stopwatch During Breaks" and "Check to See If I Turned Stopwatch Back On."

Your Inner Three-Year-Old

What is it about that age when toddlers start to ask, "Why?" It is both engaging and frustrating. Toddlers are at a stage where they are figuring out how things work. If an obliging adult is within earshot, they will pepper her with "Why?" questions.

Adult: If you're not going to finish your milk, please put your sippy cup in the refrigerator.

Toddler: Why?

Adult: That's what we do. We put it there so it won't go bad.

Toddler: Why?

Adult: Milk goes bad if you don't keep it cold.

Toddler: Why?

Adult: Well, it will start to smell bad and eventually grow mold.

Toddler: What's mold?

Adult: It's slimy black stuff that will grow in your sippy cup if you leave it out on the counter.

Toddler: Why?

Adult: Because I said so.

At this point, the adult reaches the edge of their field of knowledge and yearns to get out of the question game. But you can see how the toddler managed to get a lot of helpful explanations out of the adult before frustration set in. Your job is to play the toddler without treading on someone's sense of propriety.

this research, that I am interested in the environment in which they are working, the philosophies they follow, and the motivations that cause them to do things a certain way. I explain that I'm interested in all the tools they use. Sometimes there is a history between the participant and the group conducting the research, possibly resulting in a rocky relationship. In these cases, I do my best to make a good impression, honestly telling them everything we hope to do with the research. In one case, I excused the participant from the interview because he was just too uncomfortable with it. Do what makes sense in the situation, bearing in mind the lasting impact it will have on all those involved.

It's encouraging to know that most people love to talk about themselves and what they do. Usually the challenge is to rein them in, rather then draw them out. Even when speaking to participants from other cultures, I have found this to be true. Just make people feel comfortable, and they will tell you all sorts of things.

What Goes Through My Head While Conducting an Interview

I believe an example will be helpful at this juncture. Here's a transcript from an interview I conducted in June 2004, followed by a rundown of the steps I took in response to what the participant was saying.

Sample Moviegoer Interview

Here is a sample moviegoer interview. (Lane Becker is my former co-founder at Adaptive Path who went on to start Satisfaction Unlimited.)

Indi: Tell me about the movies you see.

Lane: I'll go see anything. That's not totally true. I won't go see anything. I have preferences, but they're all over the map. I like independent movies and the cheesy teen movies, and I always wanted to write a screen play. It would be a teen movie based on Moby Dick. I'm very excited about it. That would be hilarious. I'll go see summer blockbusters. Going to see movies is very much about setting expectations ahead of time before you see it.

Indi: How?

Lane: Going into a movie, you set the bar for what you think it will be. It puts me in contrast to Peter who decides whether the movie lives up to the high bar he set. As a result he never sees cheesy teen cheerleader movies. I will go see them and not expect anything, and some of them are brilliantly structured or well written. I don't enjoy being constantly disappointed.

Indi: What do you mean by structure?

Lane: I was a film student in college and studied screen writing. There is a standard way to roll out a Hollywood movie. It's a baseline for what a movie can be. A summer blockbuster is like paint by numbers. You know what's going to happen in the first 30 minutes and the next 60 minutes. You know how the movie is going to be. To me, it is creative when they turn it on its head and have unexpected moments, good writing, good character development, things that aren't obvious. Summer blockbusters are obvious. There is an old joke by William Goldman (who was a famous screenwriter in early 80's); the problem with Hollywood was that they always want to explain everything. He'd be sitting in a meeting after seeing the dailies and it would be a western with all these buffalo going over this big cliff, with big noise and thundering feet and some manager will say, "Hey can we add a voice-over that says, 'Here come the buffalo?'" I prefer when they don't do that, but that's okay. I get ideas out of a bad experience. (Continued on next page...)

Sample Moviegoer Interview (Continued)

Indi: What do you get?

Lane: Ideas about what not to do. Part of the fun of watching movies is thinking about construction, creative process, how people are engaged.

Indi: What else can you tell me about going to movies?

Lane: I go with Courtney; usually it's a last minute decision. We'll see a hole in our schedule. It's really rare that I will say, "Let's go see a movie next Tuesday." I'm all, "Hey, we have a hole in our schedule. Let's go see a movie!" Or two! Sometimes we have the whole afternoon and we see a couple of movies. I'll see anything. That's not true. I don't like the ridiculous romances. Some of them are so cheesy and over the top.

Indi: In what way?

Lane: Oh, "she finds love in a villa in Tuscany." I'm not going to see bad kids' movies. I have to think about that—the quality. It's what I'm interested in.

Indi: How, quality?

Lane: It's what's new. I subscribe to different mailing lists where they talk about what's coming out, and I read reviews on Fridays. There are independent reviewers that send out reviews and I pretty much know where I agree and disagree with them. So, I subscribe to mailing lists and on Friday read whole litany of reviews. Metacritic and Rotten Tomatoes. I will sit there and read the reviewers that I respect. That will effect whether or not I go see a movie. If a reviewer I respect doesn't like it, it will disincline me from going to see the movie. If they have bad reasoning or bad taste then I won't pay attention to them. The Stepford Wives—I really enjoyed the first one and the actors are ones that I really enjoy, but a bunch of reviewers said the studio trashed it and rewrote the third ending after extensive testing so I was disinclined to see it.

Right off the bat, Lane's response to my introductory question hits on several points that I jot down: What's his philosophy for choosing movies, what's the story behind the screenplay, and what's this about setting expectations? I ask about the latter first. His answer includes a reference to structure, so I ask about that, figuring that it's connected to the screenplay reference. He was a film student—how unexpected! I get a little excited that I have such a knowledgeable moviegoer here, and I immediately check myself to be sure I am not going down a path that's out of scope. Helpfully, Lane describes what he's looking for while watching a movie in terms of screenwriting and character development. I jot those down for later. Then he tells a joke, but gets right back to what he gets out of the movie experience: ideas. I ask him more about it by repeating a phrase he said. His answer touches upon two new topics: The creative process and how people are engaged. For some reason, though, I jump back to the general question. Maybe I have a gut feeling that there's more to tell me here. Indeed, Lane's answer refers to his wife and the logistics they use when deciding to see a movie. Then he jumps back to his philosophy for choosing a movie. I ask him to explain, and he replies that he's looking for quality. I ask him to explain his philosophy on quality. What comes next is an explanation of how he uses reviews to help him decide on a movie. At this point, we are only five or 10 minutes into the interview, but I have a list of additional references to explore:

- Story w/screenplay—more about film student, ideas?
- Screen writing, character development—more?
- Creative process
- How people are engaged

Thus the conversation grows on its own. All I have to do is pay attention, make sure we stay in scope, and dig down to the root reason why Lane is doing something.

When the interview is finished, you should upload the recording to a central file repository as soon as possible. Label the file using the ID number associated with the interviewee (never their name). Let the stakeholders know when a particularly good interview has been posted and refer to its ID number as the file to download if they wish to listen to it.

Ghosts and Ownership

I used to insist that a project guide join me as a silent listener on the phone for as many interviews as possible. I call this "ghosting" a call. I would actually type the transcript as the conversation unfolded, and then I would keep the project guide on the line after the interview and read through the fresh transcript. This approach offered the opportunity for me to learn a little back story from the project guide and to ask about certain industry-specific terms. By including her, I felt that the project guide would be more invested in the process and feel more like an owner of the resulting mental model diagram. If the project guide collaborated on the transcript, she would be able to explain the mental model to others in the organization. Moreover, I wanted those transcripts to be complete and correct. Reading them again immediately after the interview while the details were still fresh was the most effective way to get the transcripts in perfect shape. (I have often experienced the work of bad transcribers, who take a perfectly good recording and reduce it to a series of x's and blanks for the parts of the conversation they just don't understand. A good transcriber would never do this.)

These days I realize ghosting a live call is a tall order. Project guides and other stakeholders prefer to listen to the recordings of one or two interviews on their own time, say after the business day has drawn to a close, rather than during the middle of their busy schedule. Also, typing the transcript as the participant was speaking made paying attention to the conversation a hair more difficult. Most people don't have this skill, and hiring a real-time transcriber can be expensive. And finally, reading back through the interview right after it took place was boring to a lot of people. The value was not great enough.

In the future I will still insist that project guides ghost the calls when the topic is about an industry that might require some explanation. I can just spend five minutes on the phone with the project guide after the interview and discuss the aspects that I missed or the points that I misinterpreted.

Aliant's Basement Rooter

There was one interview I recall where the client was laughing after the participant hung up. We were talking to telecommunications consumers, and the topic was the internet service subscription they had. I was exploring how the participant went through the process of setting up the equipment. They had a basement, and placed the equipment down there. Then they kept referring to something as a "rooter." I, from California and not the northeast of Canada, thought they were talking about plumbing. I kept trying to change the subject, back to the internet equipment. My clients had been dying to tell me that "rooter" was the way they pronounced "router!"

The most common reaction I get from project guides after they ghost an interview is, "You didn't talk about our service/product/web site!" No matter how much you explain the difference between generative and evaluative research up front, it seems to take one interview for the significance of what you're doing to sink in for the project guides. Be patient; they always "get it" by the time they see the mental model.

Popular Tricks for Getting Through an Interview

I mentioned in the last chapter that I keep little notes around me to jot down a few phrases during the interview. (I would use a little piece of paper and scrawl handwritten notes—but I have difficulty reading my own grocery list, so this technique works best if you have decent handwriting skills.)

Another trick is described in *User and Task Analysis for Interaction Design* by JoAnn Hackos and Janice Redish. Indeed, they have a whole chapter on interviewing techniques, "Conducting the Site Visit—Honing Your Interviewing Skills." But on page 286 they talk about "active listening." This technique is the practice of repeating back to the interviewee what she just said as a way of asking her to explain herself in more detail.

Permission to Let Problem Participants Go

Even the best recruiters sometimes let a bad apple into the list of participants. This bad apple might be terse, difficult to draw into conversation, not interested in talking, or just plain impossible to understand. After a few exchanges where you ask a question and he answers with a single word or with streams of gibberish, you have permission to give up. Tell the person you have collected enough data, thank the person for their time, promise them the stipend, and hang up. Then go back to your recruiter and find a replacement.

I've seen people have trouble with the level of depth they should explore. It takes practice to get the right level of granularity. Most beginners skip over important details. I remember one of my first interviews where the participant mentioned that she held weekly team meetings. I thought, "Great, I know what a team meeting is," and went on to the next topic. However, when I was looking at her statement later, I realized that I didn't really know what was going on at the meeting. Were they reviewing work, stating progress, giving assignments, or maybe a few of these? I wished I had asked, "Why?"

And finally, yes, it is difficult to get the hang of asking open questions. No matter how hard you try, you will find yourself asking a leading question or two every interview. I still do this. Don't be too hard on yourself. Here is a complete list of questions that I asked during one interview, with the leading questions highlighted in red. I asked two leading questions in this interview.

Actual Interview Questions—The Good and the Bad

In your role as a consultant, what services do you provide?

Do you ever play a role in recommending enterprise software systems for your clients? *(This question is okay; he had mentioned his role already. I wanted to find out if he did this so I could ask more about it.)*

The gap you just mentioned between the business goals. Please go back and tell me what you're actually doing.

Were you involved in the definition of those business rules? *(He mentioned the definition of business rules, and I wanted to find out if he was involved and ask more about it.)*

Can you tell me where you started? What did you do when you began your first engagement?

How did you find out that they were wrong? *Were you* reading things? Talking to people? *(I should have stopped after the first part of this question.)*

How did you triage? What was that process like? What steps did you take?

How did you actually telegraph that?

What did the picture look like?

And then you said you presented that? *(Repeating what he said to encourage him to tell me more about it.)*

What were those presentations to the working folks and executives...how did they go?

So those recommendations were coming from the executives during those sessions? *(Verification of what he said.)*

What's your next step?

And were you involved in setting up the metrics? *(He mentioned setting up metrics, and I wanted to find out if he did this.)*

Actual Interview Questions—The Good and the Bad (Continued)

How did you do that?

How did you measure them?

How did you actually get that data?

So that's one of the things you're involved in now? *(Encouraging him to tell me more.)*

So you're getting some of this data out of the system and then you mention you're reporting to the executive team. *(Repeating what he said.)* How does that happen?

Let me back up a little bit. This is all about cleaning up issues from the first implementation. *(Repeating what he said.)* Is there anything else to describe what's going on?

Is this what you've been engaged in for the last 10 months? *(Verification of what he said.)*

How do you know these techniques? How do you know what will work best?

How do you keep up to date on the latest?

Can you mention a couple of examples?

Can you tell me...what are your steps in processing this information?

At what point do you refer back to them? *(Digging a little deeper.)*

It sounds like you remember all the topics you filed away.
What are some of those topic areas you have set up?

Do you help choose software? Can you go through that? *(He had mentioned choosing software, but I could have skipped this first question.)*

Who were you working for at that time?

Actual Interview Questions—The Good and the Bad (Continued)

Can you give me an idea of what your goals were and what your basic steps were for accomplishing them?

What was the step you used to find out?

At what point do you usually disengage from a client? *(He had mentioned it. I was digging a little deeper.)*

What's your last step?

How do you know about companies that have complementary technologies?

What form does the analysis take?

Once you identify who has that research, what do you do with it?

What's your next step, once you've got this research done, and you've identified new skills and people?

Plan Your International Interviews

Take advantage of telephone interviews to talk to people in distant locations and in other countries. If the organization you're doing research for is a multi-national company, this should be pretty easy to arrange. Recruiting can be done with the help of local employees, such as sales reps or customer service reps. International calls are not hard to arrange. Translators are available. Time zone differences are the biggest hurdle, really. It may mean early mornings or late evenings for you, the interviewer, depending on which continent you're on.

I have conducted international interviews at both ends of the formality spectrum. Some were laid back "call this person at their desk" contacts. Others were conducted in partnership with a local recruiting firm who also hosted the interviewee in their market research lab and provided an interview "interpreter," who did more than just translate. In the latter situation, the interviewee would arrive at an office address at the specified time, be introduced to the research offline by the interpreter, and then they would bring me into the conversation. After I rang off, they continued the conversation a few moments to tie up loose ends and took care of the stipend. Then the participant was free to be on his way. (Oddly enough, both the informal and the formal experiences were with people in Japan, where you would expect things to conform to a more formal arrangement.) I recommend following the suggestion of the local contacts you have as to the level of formality. If you have no local contacts, network with your peers to see who does. In countries where you are likely to conduct research there are usually market research firms and often usability labs and the like.

No matter which country you are dealing with, it is smart to write up two introductory documents: one for the participant and one for the interpreter.[4] These can be sent as an informal email or through other channels. They serve to make the parties involved feel more comfortable with the interview format. I have heard that some cultures regard an interview as a test. An introductory letter will help alleviate any misunderstanding or anxiety on the participant's part. With the interpreter, I will try to get in touch with her before the interviews to get familiar with the service she offers, and to familiarize her with the non-directed interview technique. So far I have encountered fabulous professionals in this vein who have had no trouble with the interview style.

Sometimes the participant wishes to speak my native language, English, in the interview. I am happy to oblige, but I still arrange for a person who could translate to be involved in the call just in case words fail us. I usually don't let the participant know that this is my safety plan, in case I injure his sense of pride. About 50% of my interviews where the participant spoke in English have been just fine. Often these participants have attended college in an English-speaking country, or they have studied English since youth.

---

4   Examples of each letter, for the participant and the interpreter, exist in the Resources section of the book site www.rosenfeldmedia.com/books/mental-models/content/resources

However, there are those other 50% of interviews where we fell back upon the translator for help, often for the rest of the interview. I feel bad for the people who wanted to exercise their speaking skills, so I do my best to make them feel respected and appreciated.

The University Effect

All my international research projects to date have studied corporate topics, such as "how do people purchase and maintain enterprise software" or "how do people in R&D, production, or material supply choose which chemicals to use?" In all of these projects, we have not found any part of the mental model that is regionally distinct. People in all countries we have studied exhibit the same mental model. I believe this is because the process they follow—the science or the best practice—was taught at the university level. Everyone employed in this manner has the same fundamental understanding of the practice. I call this "the university effect." I suspect that if I study something that is more cultural, I would find differences from country to country.

Make Transcripts

Transcripts are important. I could just leave it at that and end the chapter here, but let me explain. If you are a visual person, your work will be a hundred times easier with a transcript. If you are an aural person, you could listen to each recording and write down tasks as you hear them. Sorting out implied tasks and philosophies might not be hard for someone who is skilled, but writing or typing fast enough to keep up with the recording, or stopping it and starting it repeatedly, might become annoying to an aural person. No matter what, skipping the transcript does not make things go faster. It does not make analysis a suddenly speedier exercise. You still have the same amount of intellectual processing to do over each and every sentence.

So, make things easy on yourself and hire a transcriber. I have seen costs in the range of $100–$150 per 60-minute interview. I mentioned before that there are transcribers who are not as skilled as others. Judge the level of quality by the number of blanks in the transcript. If there are only a few, and they are labeled "garbled recording" or "two people talking at once," that is acceptable. If there are more than a few blanks because the transcriber simply doesn't understand the vocabulary, then hire someone else. Grammar, spelling, and punctuation in the transcript count for quality too. You're not the only one who will read that transcript. It might end up in the hands of an executive well after your part in the project is over. The transcript should follow standard spelling, grammar, and punctuation practices. Make sure it looks publishable.

If your transcriber is not a professional transcriber, sit down with her and go over what you need from her work. Ask her not to add words or skip words, but to type up the conversation exactly as it was recorded. Tell her that if she can't hear some part of the dialogue that is ancillary to the topic, such as the two of you commenting about a funny situation, it's not important to capture, or just ask her to capture it in brackets as [joking] or [laughter] in the transcript. (If you go over the transcript later, and you happen to add something, be sure to demarcate it with brackets, as well.) Ask her to identify the speaker, in the same manner a play is formatted. If there is a part of the recording that just doesn't make sense, ask her to mark down the time stamp, and listen to it yourself to determine the wording.

The Marlboro Man

I remember one recording where we were interviewing a man in China whose English was very good, but who had an accent. He was an engineer who designed printed circuit boards. There was one point in the conversation where he referred to the "marlboro." I thought, certainly he can't be talking about cigarettes! So I asked him more about the topic and every time he said "marlboro" I was no closer to understanding what it was. I finally gave up and moved on to another topic. The transcriber had the same problem with the recording. "Marlboro? He isn't talking about cigarettes, is he?" So we asked someone else on the team to listen to the recording. She had worked in the same general region as this man, and she unlocked the mystery. "He's saying 'motherboard.'"

With experience, conducting the interviews will become the most enjoyable part of the mental model process. It is the analysis of the transcripts that will be the greater challenge. But, as the next chapter points out, analysis can be tamed.

CHAPTER 8

Analyze the Transcripts

Comb for Tasks   132
Get Some Practice   154
Answers   162

Analyzing the transcripts has been described as both "powerful" and "painful." The powerful part is the sheer amount of knowledge you process and represent in the resulting diagram. The painful part comes in the hours spent pulling behaviors out of the transcripts and finding patterns.

This step in the process is where you focus on the detail. It is an intense period of work, and I describe several approaches to making it achievable in the "Plan Your Logistics" section at the end of Chapter 9. You begin by reading through the transcripts line-by-line, looking for phrases that represent "tasks[1]." I call that "combing." After you comb the tasks out of the first couple of transcripts, you look at them to see if they form any groups. For each additional transcript, you add tasks to existing groups or form new groups. When the grouping is complete, you create the diagram.

The process is pretty simple. The resulting diagram is powerful. The determination to plow through all the transcripts can be painful to summon, but thankfully there are personality types out there who like to do this sort of thing.

Comb for Tasks

You would think that tasks are pretty easy to spot. You studied grammar in school. You know what a verb is. A task is just what someone is doing—the verb in every sentence. Right?

Nope. First of all, we don't tend to be specific in conversation. We rely on context and generalizations. We use tone of voice, popular culture references, gestures, etc., to convey our meaning. Second, as a part of human nature, we tend to speak in terms of our intentions, reasons, opinions, and desires rather than simply what we are doing. This makes it a little more difficult to recognize a task in a transcript of someone's conversation.

---

[1] I got in the habit of using the word "task" years ago because I derived this approach from task-analysis.

What Do You Mean by "Task?"

I use the word "task" loosely. When I use the word "task," many of you in
the research field might think of "mouse clicks" and "steps to completion."
Others might think strictly in terms of "tasks and goals." Because I want just
one word to use in sentences when I'm describing this process, I use "task"
to mean actions, thoughts, feelings, philosophies, and motivations —everything that comes up when a person accomplishes something, sets something
in motion, or achieves a certain state. With this definition, more intangible
aspects of the person's experience will make it into the mental model and will
influence you to craft a better product. Tasks can be defined in many ways.
There is a movement afoot to encourage designers to expand beyond strictly
"tasks and goals."

If I say, "My colleague gave me directions. She told me to turn right, then
turn left into the driveway and park," the turning and the parking are *not*
the tasks. "I" was a tool in the hands of that colleague. She was doing the
thinking and I was just pressing pedals and turning wheels, so to speak.
My task was "follow my colleague's directions." Sure, you could argue
that pressing pedals is a task, but it is so insignificant in the scope of the
research that it should not be included.

So how do you identify a task? For the purposes of this research, you
are interested in what people are doing, thinking, and feeling—at a level
sufficiently detailed to describe intentions and approaches, but not so
detailed as to focus too narrowly and forget the overall goal of the research.
As I mention in Chapter 7 about interviewing, you are not interested in
what the person might do or hopes to do. You can *identify a task* as one of
the following.

- **Task**: a phrase stating an action or step to accomplish something
  (I walk the dogs every afternoon.)
- **Implied Task**: a not-so-clear phrase implying an action
  (Every afternoon the dogs need to go for a walk.)
- **Third-Party Task**: a phrase that mentions a task someone else does
  (My daughter walks the dogs every afternoon.)
- **Philosophy**: a phrase stating a belief or why tasks are done a certain
  way (Dogs should be walked twice a day.)
- **Feeling**: a phrase describing a person's feelings
  (Walking the dogs is my way of relaxing.)

Better yet, there are a few tests that you can use to identify something
that is *not what you're interested in.* Here are some things that sound like
behaviors but don't directly reveal how someone does something.
- **Preference**: a phrase stating a person's likes or dislikes
  (I don't like walking the dogs in the rain.)
- **Desire**: a phrase stating what the person wants
  (I am hoping to take the dogs to the new dog park when it opens.)
- **Expectation**: a belief that something will happen
  (I think everyone will clean up after their dogs at the park.)

Here are indicators that tasks, thoughts, or feelings are probably *about to
be* expressed, but this phrase itself is not the task.
- **Medium**: the tool being used to do something
  (I use biodegradable plastic baggies.)
- **Statement of Fact**: a phrase describing a circumstance
  (My dog sheds a lot.)
- **Explanation**: a phrase describing how something works
  (The retractable leash has a locking mechanism.)
- **Circumstance**: a phrase describing a situation or occurrence
  (One of the dogs ran into me while playing.)
- **Complaint**: (you ought to be able to recognize this)
  (The snap on the leash keeps pinching my finger.)

And finally, there are a couple of ways to determine whether your task is at the *incorrect level of granularity*—that is, not too specific and not too general.

- **Particular Task**: a very specific phrase explaining the little details of a task, or a task that is industry specific, where the people you are interviewing come from different industries (I open the snap at the end of the leash, and I put it through the D-ring on the dog's collar. Then I release the snap to close it.)
- **High-Level Task**: a general task statement with no detail, or detail that appears later in the conversation (I am responsible for taking care of my dogs.)

It is important to really understand this granularity issue. Go back to the defined scope of your research project. "How do prospective customers make a purchase decision?" "How do employees maintain their contractual relationship with the company?" "How do patients deal with a disease they have just been diagnosed with?" If your scope is closer to "How successfully do our users interact with our online product," you should be conducting usability tests, not making a mental model.

The Hallway Test

If I'm in doubt about whether something is the right level of granularity, I ask myself, "Would this be on someone's mind as she walks down the hall?" For example, Sylvia might be walking down the hall to her office thinking, "Shoot, I need to hurry up and write Bill a summary of my work to include in his presentation this afternoon." She would probably not be thinking "I need to open a new document on my computer." In the former sentence, you see a rich range of concepts, such as the need to hurry, the idea of supporting Bill with information, interest in the work she is summarizing, and the idea of an afternoon presentation.

Another thing to mention is the frequency or recurrence of tasks. You don't need to record the rate at which a task occurs. For purposes of the mental model, we are only interested if it occurs, not in the cycles of the task. A mental model diagram may run in somewhat chronological order, but there will be places where people will skip back or forth in the diagram, or repeat a certain task. The mental model diagram is not designed to represent these directional changes. You may safely ignore them.

Tracking Desires and Complaints

In addition to tasks, sometimes I capture desires, expectations, or complaints in a second document. Humans can't help but explain these things in conversation. As Todd Wilkens of Adaptive Path says, "There's a treasure trove of data in an interview." Sales, marketing, and research and development may learn something from reading this list. You also may want to review these lists when designing your solution to stay true to the spirit of your users.

I want to emphasize that mental model diagrams don't represent the tools or media a person is using. You're not interested in separating a task about "Tell my boss I finished the project" into "by telephone," "by email," or "in person." You only need to record the root task. Try to be tool agnostic.

Good Examples of Tasks

The following examples are a great way to illustrate the distinctions between a task and something else. You will note that I use a *verb+noun* format. I keep the verb+noun phrase succinct. And, I copy the quote from the transcript that shows the task in the first person. I keep these quotes as short as possible; otherwise, I'd end up with the whole transcript copied into my tasks. Not everyone copies quotes from the transcript; it is extra work. Folks working by hand with sticky notes may find this burdensome.

Trace a Quote Back to the Transcript

The handy thing about the quotes is that they illustrate the task more clearly than a short noun+verb phrase. When you share your work with others, or when you read the task later, you may think, "Now, what does that mean?" Feeling certain that you can figure it out, you read the quotes associated with the tasks. At this point your glee turns to frustration. "I still can't remember what was going on here."

If the quote doesn't tickle your memory, copy some of the words in the quote, open the appropriate transcript, and paste the words into the Find field. Magically you'll be whisked to the exact context of the quote and be able to read the whole conversation before and after it.

Note: If there are ellipses (...) in the quote, don't copy them. It means the quote to the left of the ellipses comes from a different place in the transcript than the quote to the right.

The verb+noun phrases all need to be in present tense. No gerunds—they just complicate things. "Targeting" should be "target." "Convincing" should be "convince." Also, use the first person, to preserve the immediacy of the tasks.

QUOTE	TASK	TYPE OF TASK
"The most important thing it does is help us to...to be more organized or better organized."	Look for Technologies to Help Company Be More Organized	Implied Task
"We are very interested in following developments in this work. It came to the attention of several of us in the department way before the business office would even consider it."	Follow Developments in My Industry	Task
"I think the first impressions do mean a lot... if you walk in the door and you're not exactly the image the person's looking for or something, you don't have a shot..."	Believe First Impressions Decide a Lot	Philosophy
"I did for my car. I have a Prius, and the navigation system and all of that.... Actually, I'm raving. I love the car. I absolutely love the car."	Rave About Hybrid Car in Online Forum	Task
"We have physical paperwork the customer is required to turn in with signature... We're depending on the customer to either get the paperwork in to us or fax or email."	Collect Signed Paperwork from Customer	Implied Task
"It must have been my first real full-paying job. First time I'd got paid. There's a certain novelty. I went to the bank to see that it had gone in."	Verify My Auto-Deposit	Implied Task
"My closer friends are married, so I think I open up to them a little bit more than my girlfriends..."	Feel More Open with Close Friends	Feeling

QUOTE	TASK	TYPE OF TASK
"Being a gadget guy, I went with the camera phone...It was only another $29. Spread that out over 36 months, that's not much."	Justify Cost of Camera Phone	Implied Task
"Spend time in the field getting raw data. Walking off poles, measuring, getting heights of where fiber is going up."	Collect Data in the Field	Task
"We ask them what they're trying to accomplish...have in their mind what they're trying to accomplish."	Ask About Design Goal	Task
"I don't think it's a very comfortable experience for either person. I mean actually trying to hook up two people who are single..."	Feel Hooking Up Two People Would be Uncomfortable for Both	Philosophy
"We have...people who are able to find out interactions between different products."	Explore How Product Interacts Among Different Products	Third-Party Task
"They decide what the table looks like, where the data gets sent."	Decide What the Table Looks Like	Third-Party Task
"I would be comfortable meeting somebody from the Internet in a group setting...for safety reasons... me it's trying to go out in groups, not one-on-one..."	Feel More Secure with Friends in Group Environment	Feeling
"Father told me, pay yourself first. Take money off the top and set it aside ...My father is very smart when it comes to investments. He's helped me set things on the right course."	Learn Financial Behavior from My Family	Implied Task

As you can see, with the implied tasks you are required to interpret a little. This is the "art" of the process. You will find it easier if you ask yourself, "What is this person really trying to do?" Get at the *root task*. Apply the hallway test. Here's an example from the transcript of an interview with a woman in Japan.

> "Because of the corporate security policy, I cannot use my Palm at work. I have the Outlook schedule synchronized to my Palm as I put my Palm into the cradle at home, although I don't update it so often...because [I have a] paper-based organizer, the notebook, schedule book."

Is there a task here? Why is she using a paper-based organizer in addition to the PDA? I looked back at the transcript and read that she uses paper in the work meetings she attends because it's frowned upon to use the PDA. The root task she's really talking about is "Synchronize Paper Notes with Electronic Schedule."

Task Examples That Need Work

When you first start out, it's really difficult to sift tasks out of a transcript. Most often you will see tasks in every sentence. Chiara Fox, an Information Architect at Adaptive Path, says, "Determining what is fact and what is a task is harder than it seems."

The following examples illustrate tasks that someone pulled from a transcript, with my explanation of why they are not tasks and, if possible, where the analyst could find a task in each quote. These examples were pulled from hundreds of email conversations I have had with people over the years. (My thanks to those who contributed so generously!)

QUOTE	TASK	TYPE OF TASK
"That's why we made the decision to purchase this product to provide the framework to build our data warehouse."	Provide a Framework	High-Level Task

This phrase is a goal, but not quite a task. The interviewer (yours truly) didn't drill down into the detail yet. This happens a lot in conversation. The first question I ask elicits a murky response, so I delve deeper to get to the more detailed tasks. If you notice this, don't bother pulling tasks from the first answer because they'll be repeated in much more depth in the follow-up questions. You can skip this one entirely.

QUOTE	TASK	TYPE OF TASK
"The marketing dept has to come up with new ways of persuading advertisers to place ads with us."	Persuade Advertisers to Place Ads	Particular Task

This task is very industry-specific within the scope of the research. We didn't interview any other newspaper folks. If we record it, it will fall in a tower and mental space all by itself. Concentrate instead on the scope of the research. In this case, the scope was "How do people purchase, install, and maintain our products?" In addition, this is a third-party task. It is the folks in the marketing department who are creating packages to sell to advertisers, targeted to certain audiences. This guy, himself, is just describing what these other folks are doing. I would ignore this task.

QUOTE	TASK	TYPE OF TASK
"...marketing to be able to correlate Herald data with external data of some kind."	Provide Marketing with Information	Vague Noun

Here the noun, "information," is kind of vague. Ask yourself if there is anything in context that would allow you to clarify it. Here is a larger snippet from the transcript: "Yeah, I think it's important also for marketing to be able to correlate Herald data with external data of some kind...One example is a company called InfoUSA that markets telephone numbers and addresses across the US. I would take a sample size..." When I read the context of the quote, I came up with this task instead: "Correlate

Data among Internal & External Sources." I discovered the root task and changed the verb, too. This is a third-party task.

QUOTE	TASK	TYPE OF TASK
"Now we have people doing the laborious work of number crunching and taking multiple spreadsheets and adding them together by hand."	Manual Number Crunching	Missing Verb

The topic at this point in the transcript was about a team of folks figuring out the answers to questions the hard, old-fashioned way, rather than just making queries to a database. The root task here is that they got annoyed enough with the hard labor that they decided to buy a data warehouse. "Decide to Warehouse Data to Avoid Laborious Number Crunching" is a better way to portray it.

QUOTE	TASK	TYPE OF TASK
"We brought in a few people to make presentations."	Bring People in for Presentations	Vague Noun

It's okay to replace the word "people" with the word "vendors" so that it has a little more meaning. This participant does use the word "vendor" later.

QUOTE	TASK	TYPE OF TASK
"We started bringing the information together about these things."	Collect Information	High-Level Task

The quote goes on to describe the various ways they achieved this goal. We can ignore the "bringing the information together" part because it's described in more detail as the interviewer elicited more elaboration.

QUOTE	TASK	TYPE OF TASK
"I read industry magazines. IT magazines have been mentioning business intelligence. I follow developments in this work. It came to the attention of several of us in the IT dept way before the business office would even consider it."	Read Industry Magazines	Vague Verb

The essence of what he is saying is that he and his team became aware of a technology before the rest of their company, and therefore had to evangelize, persuade, and educate the business office as to the benefits. He follows developments in the IT world through these magazines. This task can be more powerful if worded like this: "Pursue Concepts from Industry Magazines." He does more than just read.

QUOTE	TASK	TYPE OF TASK
"We got the marketing department's main person to see the demo. She instantly saw the power. It didn't take her long to catch on to what this could do for her."	Demo Possible Solutions	Vague Verb

This task doesn't seem to match the quote. It's a well-formed task, but the quote says that the person brought in the director of the marketing department to "Explain the Power of a Solution to Peer."

QUOTE	TASK	TYPE OF TASK
"More recently we're turning our attention to the AJAX aspect of it as well."	Turn Attention to AJAX	High-Level Task

Here he is speaking at a high level about his job responsibilities in the first minute of the interview. This is a high-level task. Often, high-level tasks appear in the first paragraph of the transcript, and concrete detail appears later. You can ignore this task.

QUOTE	TASK	TYPE OF TASK
"Many years ago we decided we don't want reporting on our mainframe every night because this reduces our CPU usage."	Optimize System Usage	Wrong Verb

This is a very nicely worded task. However, the quote does not talk about optimizing. It talks about reducing. I'd label this one "Reduce CPU Usage."

QUOTE	TASK	TYPE OF TASK
"We look at size of the databases and see which ones need more space. We monitor what space they take up on our limited resources."	Continuously Monitor Systems	Extra Adverb, Vague Noun

This phrase could be shortened to "Monitor" without losing anything. Instead of "systems" you could say "Available Disk Space" to be clearer about the root task.

QUOTE	TASK	TYPE OF TASK
"We can tell what percentage is filled with data and indices and such and what percentage is free. We get those reports every week and see how it's growing and anticipate."	Anticipate Expansion Needs	Vague Verb

This phrase has a bit of a weak verb. "Anticipate" is used correctly here, but it does not connote a particular activity. "Plan for Data Storage Expansion Needs" would be better, since he uses the word "plan" somewhere else in context.

QUOTE	TASK	TYPE OF TASK
"We plan months in advance."	Plan for Software Upgrades in Advance	Extra Words

This task could do without "in Advance" because it's implied by "Plan."

QUOTE	TASK	TYPE OF TASK
"We have various people who design the data warehouse. I'm not the only one who keeps track of it; we have a little committee."	Implement DW Directions	High-Level Task

This task goes a bit beyond what he is saying in the quote. He's just saying that he and the committee both track the data table design, so "Track Table Design" could be one task. You could add a third-party task saying, "Design the Data Warehouse."

Red-Flag Vague Verbs

There are verbs which are too vague to help define a task. Here is a list I have collected over the years. Avoid using these verbs if you can. There is always a better alternative when you think about the root task under the surface statement.

- *Consider* "Consider file size when emailing files" ➜ "Reduce file size when emailing files"
- *Deal With* "Deal with problems as they come up" ➜ "Understand an error" and "Research a solution" and "Fix a problem"
- *Find* "Find it difficult to use at work" ➜ "experience difficulty using product"
- *Get* "Get an email" ➜ "Check my inbox for email"
- *Have* "Have problem uploading file" ➜ "Ask friend for help uploading file"
- *Let* "Let employee give presentation" ➜ "Approve employee request to give presentation"
- *Manage* "Manage my contact list" ➜ "Add new contacts to my list" and "Edit contact information that has changed"
- *Plan* "Plan my project" ➜ "List elements that I want to include in my project" and "Schedule my project on a timeline"
- *Read* "Read the report" ➜ "Study the report for possible improvements"
- *Receive* "Receive directive from management" ➜ "Learn about a directive from management"
- *Use* "Use dial-up before high speed" is a statement of fact, not a task.
- *Want* "Want to learn" is a desire, not a task

Compound Tasks

When I help folks who are starting out with combing, I see a tendency to either go too detailed or too complex with tasks. The latter tends to show up as compound noun phrases or tasks with the word "and" in them. You want your tasks to be as atomic as possible. If you find a complex task, break it up. Consider the following example.

QUOTE	TASK	TYPE OF TASK
"The OEM application was attractive in terms of its design-capability. It doesn't require a database administrator to use. If a client has a problem they can ship us the database and we can examine it and find what went wrong."	Consider OEM Application to Be Able to Ship and Examine Database	Compound Task

This makes three tasks:

QUOTE	TASK	TYPE OF TASK
"The OEM application was attractive in terms of its design-capability. It doesn't require a database administrator to use."	Decide on OEM Application for Ease of Use	Implied Task
"If a client has a problem they can ship us the database..."	Ask Client to Ship Us the Database	Implied Task
"...we can examine it and find what went wrong."	Examine Client Database for Problem	Task

The idea is to simplify to the root tasks. Another example, "Run and Maintain Servers" could just be "Run Servers," or "Maintain Servers," or "Keep Servers Running," which is the most powerful way to convey the meaning. If you find that you are using "and" between verbs, it's a flag that you should re-examine the task description. Split the quote into component parts and make two tasks, each with one verb. Or, as is the case here, double-check the transcript. Here the speaker is really just repeating himself. You could argue that "maintain" is slightly different than "run." However, later in the transcript he speaks more specifically about updating the servers as a part of maintenance. So we can just use the "keep 'em running" interpretation here.

In another example task, "Decide to Get External CD Drive to Replace Broken Drive versus Buying New Computer" expresses several things to me, for example: decide to get CD drive; replace broken CD drive; compare replacing drive to cost of new computer. And then I read the quote and she's only *planning* to get an external CD drive. It's not even something she's done yet. So, after much study, I came to the conclusion that her task was "Compare Replacing Drive to Cost of New Computer," because that's something she has actually done. It was not a compound task after all.

In the task, "Keep Up to Date on Software," the quote from the transcript is "We'll watch the emergency bug reports...and we'll pick out patches." This task is really two root tasks: "Watch Bug Reports" and "Pick Patches from Bug Reports."

A final example is "Watch For and Select Software Patches." The keyword "and" appears here. This phrase breaks into two discrete tasks: "Read Emergency Bug Reports" and "Select Patches."

Formatting Tasks

You can capture the tasks in any way you want, using, for example, Excel, Word, a database, or a wall of sticky notes. Some teams will work well rubbing elbows in the conference room, reading aloud and furiously scribbling on bits of paper. Other teams work remotely, passing files around. Some teams have both active and passive members, so working on an electronic document projected on a screen in a conference room works well. Choose a mode that suits your team, and don't be afraid to switch modes midstream.

For teams that use electronic files, I have a few suggestions about file format. In an ideal world, you will copy a quote from the transcript that supports each task you write, as well as some data about the participant who was speaking. Strictly speaking, it isn't necessary to write down anything other than the task, but the extra data helps with later analysis. I use either a spreadsheet or a word processor to store my tasks. I have heard of people using database files, as well. In my spreadsheets and

International Teammates

Happily we live in an age where we can work with geographically separated team members. With email, instant messages, conference calls, real-time internet file sharing, and document repositories, we can interact freely with remote peers. Time zones, language, and cultural differences remain a problem to be solved, though. Multi-national corporations have each developed a method to work with teams in different countries that include ideas like cycling the meeting time so that it occurs at a different hour each session, so one party does not always suffer the early-morning or late-night conference call.

It's nice to meet your co-workers face-to-face to establish a relationship, but it's not necessary. If you can get together in one location at least once, I highly recommend it. But if that is impossible because of budget constraints, the environmental cost of flying, or politics, then develop connections and trust via phone and other communication media. There are a few people I have worked with in the past who I never met face to face, and the projects were a great success.

documents, I list one task per line, with the associated quote and data on that same line (see example in Figure 8.1).

For teams that use sticky notes, you might consider trying to capture data about the participants behind a task with some sort of notation on each sticky note. For example, note the participant ID number on the sticky note, and maybe other things you think might delineate differences in behavior, such as geographic location or role.

In this spreadsheet example, the Atomic Task column contains the task names that you write, based on the quotes you paste into the Quote column. The ID# identifies which participant said this, without propagating their names throughout the document. This protects their

Atomic Task	ID#	Aud Seg	Loc	Quote
Pursue BI Concepts from Industry Magazines	101A	PSN	NA	Because I read industry magazines. IT magazines have been mentioning business intelligence.
Read industry magazines, trade press	104	PrSOE M T	NA	I have lots of biz magazines... industry, trade press.
Subscribe to and read vertical industry publications	112	PRESO T	NA	Day to day through either weekly or daily newsletters, which are either business journals or industry tracking... as well as a couple of specific telecom
Read neutral trade newsletter vendor comparison reports	104	PrSOE M T	NA	The trade newsletters would also do comparisons so I had an idea of what box compares with that box. I used neutral sorts of sources
Use industry research newsletters to learn	115	PRENB	EU	[Are these research things Forester and Gartner?] Yes. Newsletters.
Follow developments in DW industry	101A	PSN	NA	very interested in and following developments in [data warehousing] work. Came to attention to several of us in the IT dept way before the business
Actively survey and consider new technology from business and tech side	104	PrSOE M T	NA	I survey technology and have an opinion about it, both from a technological side and from a business side
Research technologies and products to use	115	PRENB	EU	dealing with research to find exactly what technologies and products we need
Keep up with the IT industry to survive in IT	101A	PSN	NA	There is no way to survive in the IT world without reading something
Subscribe to and read vertical industry publications	112	PRESO T	NA	General IT industry
Learn from tech team's expertise for latest industry developments	105	PrSOE M B	NA	our technical people, their trade magazines, internet searches or reviews... industry group meetings... One of the tech people might say, "This

FIGURE 8.1

Tasks recorded in spreadsheet format. (Ignore any grammatical errors in this raw data. They are a result of transcribing spoken conversation.)

privacy. The Audience Segment is represented by an abbreviation. The Location depends upon the scope of your research. In this example, NA stands for North America, and EU stands for Europe.

In the second row, you see that the quote has ellipses (...) in the middle of it. I use this to string together a few parts of the transcript into one statement that supports the task. The ellipses signify that the words "magazines" and "industry" are not contiguous in the transcript. If I want to find this quote in the transcript file, I can do a search on any phrase except one containing the ellipses. You see that I use ellipses in rows three and eleven, as well.

It is important to have only *one line per voice per task*. You don't want two rows representing one task when both rows are the same person talking. If the person repeats himself later in the transcript again and again, you can copy all the different quotes and put them in one cell, with ellipses between the quotes to indicate that the quotes were taken from separate areas of the transcript.

In this example (as shown in table in Figure 8.2), you can see several different voices supporting the same task. "Get New Product Concept from Marketing" has two voices, both from audience segment S, and both from Japan. These two voices represent two different people. (Note that I did not put a column for ID# in this project, to suit the client's privacy requirements. I kept track of the IDs in my head.) For "Collect New Product Needs from Customers," there are five supporting voices, two from Japan, one from Spain, one from Great Britain, and one from Germany. Five people spoke about how they collect product ideas from customers. Note that I do not rewrite the name of the task on each of these supporting lines, but you certainly can. Alternately, you can put a "ditto" mark (") in the empty cells to indicate that they are the same as the cell above.

On the last line, you also see the use of an asterisk (*) in the Audience Segment column. This asterisk is my way of noting that the speaker, although he belongs to the BC audience segment, was speaking about someone else who belongs to the S audience segment. This notation is a good way to denote third-party tasks.

Atomic Tasks	A	Ctry	Supporting Quotes
Get New Product Concept from Marketing	S	JPN	Every day the research lab receives people from marketing and sales. Depends on the topic, he spends 30 minutes to 2 hours a day with the marketing and sales.
	S	JPN	Marketing colleagues announce to my development people that this concept is highly evaluated by the consumer... This is a typical starting point of a new product, of a local product.
Collect New Product Needs from Customers	S	JPN	I meet customers, collect the needs from the customers... talk with their purchasing department people or technical people... After 5pm, we have more social occasions, to collect the needs, while drinking.
	R	ESP	If one particular market segment has more than 10 customers asking for the same product we don't have, and it looks to be a product we can make, we need to get in there.
	S	JPN	Also have to cover other themes that come from customers... Customer's first contact actually is Lintec, to tell the company what the new needs are. They come to our company: "Can you solve this?"... As for spot-problem solving for the customer, they receive 300 requests per month.
	S	GBR	We're the central development work. If a customer has an application.
	S	DEU	One aspect is what the customer of our company requires, in some cases the customer tells us this is what we need... sales force direct contact with our customers, in constant dialog... getting constant feedback about new product development and what the customer requires.
Get Specification from Government	BC* S	CHN	The government they say, "We want ABC" and we put ABC together.

FIGURE 8.2.

Tasks recorded in word processor format, as a table. (Ignore any grammatical errors in this raw data. They are a result of transcribing spoken conversation.)

One thing to mention is that a quote should support one task. If it seems to support multiple tasks, rip it apart so the sections of the quote each support a different task. If you find yourself wanting to use the same quote for two tasks, that's an indication that the root task is unclear. Try to split the quote so that each phrase supports a distinct task. If that really makes the task context unclear, try using the same quote for two tasks, but style the different part of the quote you mean for each task as bold.

Try not to add any words to the quote. You want the quote to *literally* match the transcript. When reviewing this with a person unfamiliar with the research, you will want to highlight phrases and do a search in the transcript to show the person the context of the conversation. If there's something added to the quote in the file, you will not be able to find it in the transcript.

The Powerful Verb

As I mention in Chapter 1, when I first meet a team that I'm going to work with, I ask them, "What benefit will this project have for your users?" I get answers such as, "It will increase our customer satisfaction scores," or "This will allow us to publish more." It has been drilled into our heads to think of the customer first, but evidently lots of people have a hard time articulating the customer's perspective, as indicated by these kinds of answers.

You already have noticed that I use verbs as the tool to expand the team's perspective. When forced away from nouns, team members are required to come up with descriptive words that illuminate what the user is doing. Often these verbs are the user's own vocabulary. Ferreting out the right verb to describe a task exercises a new cerebral muscle. You think harder about how the user sees it. Over time, you become more adept at thinking in verbs. By default, you also have switched to a deeper understanding of what the customer is trying to do.

I ask the team to stick with verbs until they are ready to start labeling sections of the navigation. At that point, it's safe to return to "Noun Land." Until then, there is great benefit in thinking exactly like a user.

> ### Scholarly Phrases
>
> I've noticed a funny thing: we "information workers" tend to translate things into scholarly phrases. I'll write "Inform Manager of Incident" when I mean "Tell My Boss I Spilled Coffee on My Keyboard," or "Implement Instructional Strategy" when I mean "Teach Class the Way I Planned." I catch myself doing this all the time. Keep your eye open for translations like these and try to return the phrase to its native, colloquial language.

Get Some Practice

In the following list, are these tasks or not? Your job is to identify the imposters. Check box per statement; if you're really good, you'll be able to say which type of non-task the imposters are.

1. Make sure customs allows equipment to be brought in

SOMETHING YOU WOULD RECORD	SOMETHING YOU WOULDN'T RECORD
✸ task	✸ preference
✸ implied task	✸ desire
✸ third-party task	✸ expectation
✸ philosophy	✸ medium
✸ emotion	✸ statement of fact
	✸ explanation
	✸ circumstance
	✸ complaint
	✸ particular task
	✸ high-level task

2. Be familiar with moving

SOMETHING YOU WOULD RECORD	SOMETHING YOU WOULDN'T RECORD
✹ task	✹ preference
✹ implied task	✹ desire
✹ third-party task	✹ expectation
✹ philosophy	✹ medium
✹ emotion	✹ statement of fact
	✹ explanation
	✹ circumstance
	✹ complaint
	✹ particular task
	✹ high-level task

3. Know process from past experience

SOMETHING YOU WOULD RECORD	SOMETHING YOU WOULDN'T RECORD
✹ task	✹ preference
✹ implied task	✹ desire
✹ third-party task	✹ expectation
✹ philosophy	✹ medium
✹ emotion	✹ statement of fact
	✹ explanation
	✹ circumstance
	✹ complaint
	✹ particular task
	✹ high-level task

4. Get health coverage quickly because of cost

SOMETHING YOU WOULD RECORD	SOMETHING YOU WOULDN'T RECORD
✱ task	✱ preference
✱ implied task	✱ desire
✱ third-party task	✱ expectation
✱ philosophy	✱ medium
✱ emotion	✱ statement of fact
	✱ explanation
	✱ circumstance
	✱ complaint
	✱ particular task
	✱ high-level task

5. Put money away

SOMETHING YOU WOULD RECORD	SOMETHING YOU WOULDN'T RECORD
✱ task	✱ preference
✱ implied task	✱ desire
✱ third-party task	✱ expectation
✱ philosophy	✱ medium
✱ emotion	✱ statement of fact
	✱ explanation
	✱ circumstance
	✱ complaint
	✱ particular task
	✱ high-level task

6. Experience difficulty following training

SOMETHING YOU WOULD RECORD
* task
* implied task
* third-party task
* philosophy
* emotion

SOMETHING YOU WOULDN'T RECORD
* preference
* desire
* expectation
* medium
* statement of fact
* explanation
* circumstance
* complaint
* particular task
* high-level task

7. Want to get stock options

SOMETHING YOU WOULD RECORD
* task
* implied task
* third-party task
* philosophy
* emotion

SOMETHING YOU WOULDN'T RECORD
* preference
* desire
* expectation
* medium
* statement of fact
* explanation
* circumstance
* complaint
* particular task
* high-level task

8. Get boss involved to make decisions

SOMETHING YOU WOULD RECORD
* task
* implied task
* third-party task
* philosophy
* emotion

SOMETHING YOU WOULDN'T RECORD
* preference
* desire
* expectation
* medium
* statement of fact
* explanation
* circumstance
* complaint
* particular task
* high-level task

9. Have lean global team for efficiency

SOMETHING YOU WOULD RECORD
* task
* implied task
* third-party task
* philosophy
* emotion

SOMETHING YOU WOULDN'T RECORD
* preference
* desire
* expectation
* medium
* statement of fact
* explanation
* circumstance
* complaint
* particular task
* high-level task

10. Go to court

SOMETHING YOU WOULD RECORD
* task
* implied task
* third-party task
* philosophy
* emotion

SOMETHING YOU WOULDN'T RECORD
* preference
* desire
* expectation
* medium
* statement of fact
* explanation
* circumstance
* complaint
* particular task
* high-level task

11. Distrust sales reps

SOMETHING YOU WOULD RECORD
* task
* implied task
* third-party task
* philosophy
* emotion

SOMETHING YOU WOULDN'T RECORD
* preference
* desire
* expectation
* medium
* statement of fact
* explanation
* circumstance
* complaint
* particular task
* high-level task

13. Help people get information

SOMETHING YOU WOULD RECORD
* task
* implied task
* third-party task
* philosophy
* emotion

SOMETHING YOU WOULDN'T RECORD
* preference
* desire
* expectation
* medium
* statement of fact
* explanation
* circumstance
* complaint
* particular task
* high-level task

14. Pick a free time to respond to email

SOMETHING YOU WOULD RECORD
* task
* implied task
* third-party task
* philosophy
* emotion

SOMETHING YOU WOULDN'T RECORD
* preference
* desire
* expectation
* medium
* statement of fact
* explanation
* circumstance
* complaint
* particular task
* high-level task

15. Save my work file

SOMETHING YOU WOULD RECORD
* task
* implied task
* third-party task
* philosophy
* emotion

SOMETHING YOU WOULDN'T RECORD
* preference
* desire
* expectation
* medium
* statement of fact
* explanation
* circumstance
* complaint
* particular task
* high-level task

16. See ideas in magazines

SOMETHING YOU WOULD RECORD
* task
* implied task
* third-party task
* philosophy
* emotion

SOMETHING YOU WOULDN'T RECORD
* preference
* desire
* expectation
* medium
* statement of fact
* explanation
* circumstance
* complaint
* particular task
* high-level task

Answers

1. Make sure customs allows equipment to be brought in = *task*
2. Be familiar with moving = *statement of fact*
3. Know process from past experience = *explanation*
4. Get health coverage quickly because of cost = *philosophy*
5. Put money away = *task*
6. Experience difficulty following training = *statement of fact*
7. Want to get stock options = *desire*
8. Get boss involved to make the decision = *task*
9. Have lean global team for efficiency = *philosophy*
10. Go to court = *high-level task*
11. Distrust sales reps = *emotion*
12. Avoid viruses = *high-level task*
13. Help people get information = *high-level task*
14. Pick a free time to respond to email = *task*
15. Save my work file = *particular task*
16. See ideas in magazines = *medium*

CHAPTER 9

Look for Patterns

Group Tasks into Patterns   164
Plan Your Logistics   189
Congratulate Yourself   195

The second half of the analysis process involves grouping tasks by affinity. As you add more tasks, you build the groups into ever-larger sets. It is a simple process, but it will take days to complete. (If it's any help, repeat the athlete's mantra, "The pain you feel is weakness leaving your body.")

Group Tasks into Patterns

Work from the bottom up. In other words, let the tasks find their own patterns, rather than try to force them into a pre-existing set of groups that you have in mind. Start at the smallest level of granularity and let things build from there. Seriously, try not to impose a structure of your own, accidentally or on purpose. Let the tasks speak for themselves. The opportunity to create a new structure based on the smallest building blocks allows you to capture a more honest reading of your user's tasks.

Here's a picture of what not to do (Figure 9.1). You have a bunch of tasks, but you've already created some empty boxes to put them in. Where would you put "Decide When to Hold Meetings in Each City" or "Figure Out which City to Spend the Night In?" Forcing these tasks into existing boxes breaks the integrity of the model.

The Order of Steps

Once you have two or more transcripts, begin combining the tasks you have discovered. What tasks are similar to each other? When the tasks are truly root tasks, it's fairly easy to determine which ones are alike. Those who watched *Sesame Street* on television learned how to do this at a tender age. Remember the game "Which of these things is not like the other?"

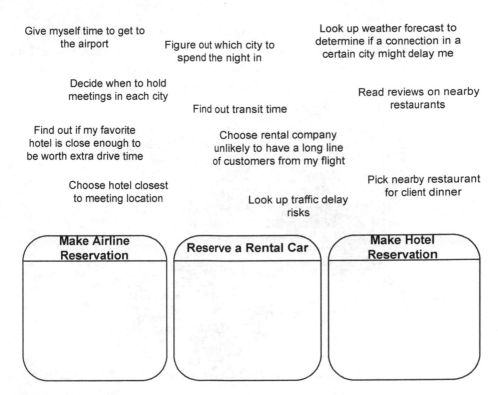

Give myself time to get to the airport

Figure out which city to spend the night in

Look up weather forecast to determine if a connection in a certain city might delay me

Decide when to hold meetings in each city

Read reviews on nearby restaurants

Find out transit time

Find out if my favorite hotel is close enough to be worth extra drive time

Choose rental company unlikely to have a long line of customers from my flight

Choose hotel closest to meeting location

Pick nearby restaurant for client dinner

Look up traffic delay risks

Make Airline Reservation

Reserve a Rental Car

Make Hotel Reservation

FIGURE 9.1.
Imposing a structure on your tasks? Bad. No. Don't do this.

I still love this game. It shows that things aren't always black and white. There can be many answers, and one of the answers is the best answer. For instance, if you follow the link in the caption (as shown in Figure 9.2, if it's still valid) and keep playing the game, you'll come across a set of four images that looks like this:

FIGURE 9.2.
Sesame Street affinity game (tinyurl.com/2cnhah).

What thought process did you go through when you saw the pictures? Your thinking might have been similar to mine. The sailboat and the balloon both have cloth. Many small aircraft are covered in cloth then painted, so the helicopter could be the answer. But the aircraft shown here looks like a fighter jet, which has metal skin. The boat, helicopter, and balloon are all colorful in the pictures, so that could mean the jet could be the answer. The boat, the jet, and the helicopter all could be piloted by someone named the captain, so the balloon could be the answer. But since the balloon is technically a vessel, you also could call the person controlling it a captain. I could compare them by the noise they make, by the way they use the wind to travel, or by the way they move vertically through the air. However, the right answer, according to *Sesame Street*, is that the jet, helicopter, and balloon all travel through the air, and the boat travels on water. There are a lot of answers, but this is the best answer.

This is the same process you will go through when comparing tasks and then grouping them. You'll see similarities between tasks and start to pile them together. There will be some tasks that are so similar that they really mean the same thing, like in the following quotes from three different transcripts:

> "My friend really likes talking about the movie, so we'll end up talking."

> "...talk with afterwards... What we like or dislike... Whether it was good"

> "I'm interested in hearing it after."

These quotes combine to form one atomic task called "Discuss Film Directly After Viewing." You saw another example of combining voices in Figure 8.2.

Discuss Film Directly After Viewing

> "My friend really likes talking about the movie, so we'll end up talking."

> "...talk with afterwards... What we like or dislike... Whether it was good"

> "I'm interested in hearing it after."

Frequently, many atomic tasks combine to form a new layer, which is the task level. In this example I combine our first atomic task with another atomic task called "Discuss Films with Friends Later" to make a task called

"Discuss Film Afterwards." Discussing a film later, in the scope of this particular research, is similar enough to discussing films directly after viewing to combine them. Both the atomic tasks illustrate roughly the same process.

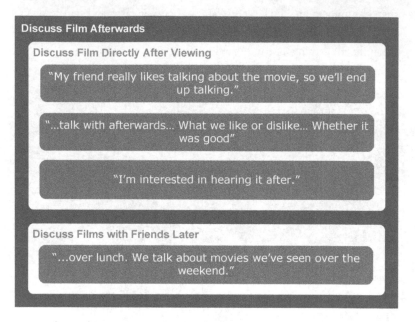

When you combine "Discuss Film Afterwards" with other tasks like "Go Somewhere to Sit and Discuss Film," "Learn Craft from Discussion of Unusual Points," "Discuss Interpretation of Book," and "Ask Strangers Their Opinion After a Film," then you have a tower. Each of these tasks is a different process. Discussing the interpretation of a book into film is different than asking strangers their opinion of a film. A tower is the next level of granularity up. Towers contain tasks that are conceptually related in nature, but are not the same. The tower itself represents a more general level of thinking. In the example, the tasks I list combine to form a tower called (can you guess?) "Discuss the Film."

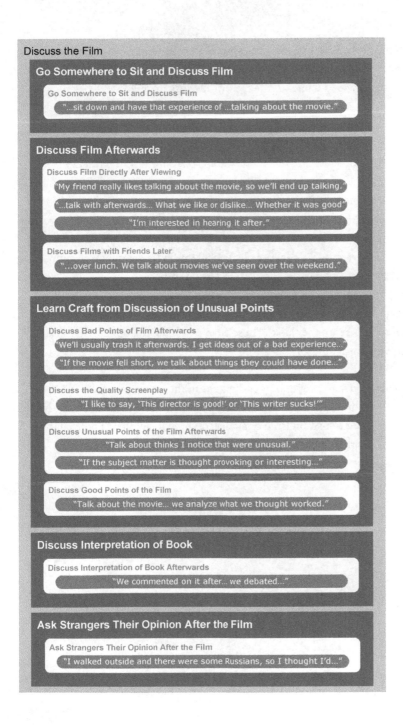

Discuss the Film

Go Somewhere to Sit and Discuss Film

Go Somewhere to Sit and Discuss Film

"...sit down and have that experience of ...talking about the movie."

Discuss Film Afterwards

Discuss Film Directly After Viewing

"My friend really likes talking about the movie, so we'll end up talking."

"...talk with afterwards... What we like or dislike... Whether it was good"

"I'm interested in hearing it after."

Discuss Films with Friends Later

"...over lunch. We talk about movies we've seen over the weekend."

Learn Craft from Discussion of Unusual Points

Discuss Bad Points of Film Afterwards

"We'll usually trash it afterwards. I get ideas out of a bad experience..."

"If the movie fell short, we talk about things they could have done..."

Discuss the Quality Screenplay

"I like to say, 'This director is good!' or 'This writer sucks!'"

Discuss Unusual Points of the Film Afterwards

"Talk about thinks I notice that were unusual."

"If the subject matter is thought provoking or interesting..."

Discuss Good Points of the Film

"Talk about the movie... we analyze what we thought worked."

Discuss Interpretation of Book

Discuss Interpretation of Book Afterwards

"We commented on it after... we debated..."

Ask Strangers Their Opinion After the Film

Ask Strangers Their Opinion After the Film

"I walked outside and there were some Russians, so I thought I'd..."

You can see how this tower represents a less-granular concept than the tasks themselves. You will want to pull vocabulary from the task labels to come up with a name for the tower, "Discuss the Film."

Look at the table again and notice how the tasks are the main element. Some tasks are made up of just one atomic building block. Others are made up of a few atomic tasks. And some atomic tasks represent several voices. If you were building this table from the top down, this structure would be too random and chaotic to handle. But since you build the table from the bottom up, the structure is a natural byproduct of the data you are working with. There is nothing strange or random about it. If you go out and interview additional people, you might find more quotes to support the singletons like "Discuss Interpretation of Book." For example, you might hear about expressing dismay over missing scenes, feeling upset that the message was warped, or praising devices by which the director inferred parts of the book. These atomic tasks all roll up into "Discuss Interpretation of Book."

When you start seeing bigger divisions in the tasks and towers, then you can start dividing them into mental spaces. Look for divisions in the towers that mark a transition from one frame of mind to another. These clusters of towers that make a mental space all have to do with the same goal or something that might happen around the same time.

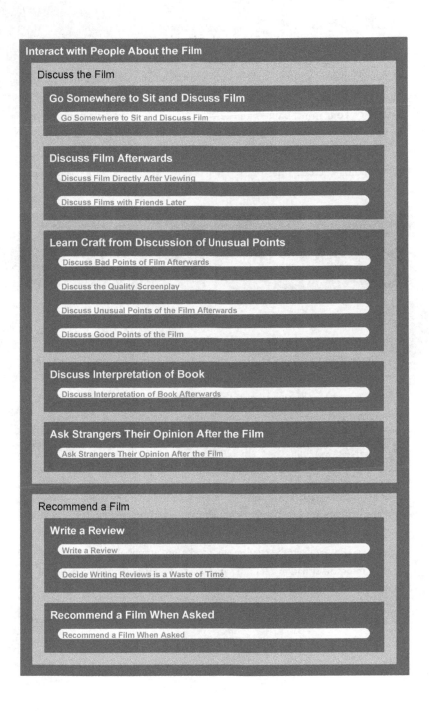

Interact with People About the Film

Discuss the Film

Go Somewhere to Sit and Discuss Film

Go Somewhere to Sit and Discuss Film

Discuss Film Afterwards

Discuss Film Directly After Viewing

Discuss Films with Friends Later

Learn Craft from Discussion of Unusual Points

Discuss Bad Points of Film Afterwards

Discuss the Quality Screenplay

Discuss Unusual Points of the Film Afterwards

Discuss Good Points of the Film

Discuss Interpretation of Book

Discuss Interpretation of Book Afterwards

Ask Strangers Their Opinion After the Film

Ask Strangers Their Opinion After the Film

Recommend a Film

Write a Review

Write a Review

Decide Writing Reviews is a Waste of Time

Recommend a Film When Asked

Recommend a Film When Asked

You can separate two mental spaces by thinking of it as the pause between the main steps people are using to accomplish their goals. Imagine you are deeply absorbed in watching your favorite television program and your significant other or roommate comes in and asks, "Did you take out the garbage?" Your mind pauses a little while it switches gears from watching the program to understanding what you were asked and formulating an answer. This is similar to how mental spaces differ from one another. The person doing the tasks has paused and is now on to something else. Here are some other mental spaces in the example I've been showing:

Watch the Film	Identify with the Film	Interact with People About the Film	Follow the Industry

You can see how there is a pause between these steps. There is also a chronological order to these tasks, but that is happenstance. The diagram that results from this process is not designed to depict directional movement or cycling. Go ahead and put your mental spaces in a likely order, but don't pull your hair out about chicken-and-egg situations where one mental space could come before *and* after another.

What If a Task Belongs in Many Places?

Don't copy and paste one task into different groups. If it seems like a task really belongs in many places, this indicates the task is compound. The task signifies a few meanings and needs to be broken down into root tasks. Break it down and then sort the resulting root tasks into groups. You will probably encounter this situation a lot. I do. Just keep refining your thinking until you really hit the root task.

Shifting Patterns

Have you ever played with the computer game, Conway's Game of Life?[1]
It is a game where "cells" in a grid live or die depending on how many
of their neighbors are alive or dead, thus "giv[ing] rise to...structures."[2]
The patterns displayed on the grid shift and evolve over time based on the
execution of the rules (Figure 9.3).

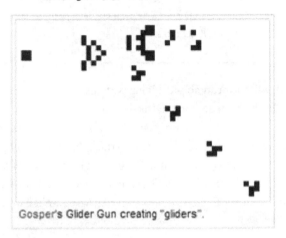

Gosper's Glider Gun creating "gliders".

FIGURE 9.3.
Image from Wikipedia Game of Life page.

Your tasks, groups, and mental spaces will also shift and evolve over time
based on the decisions you make about similarities and root meanings
during analysis. Often one mental space or tower will get too big and you'll
notice sub-groups in it and parse them out. Sometimes one of these sub-
groups gets added to an existing group somewhere else. What happens
to me a lot is that I initially sort things into two piles (among all the
other piles in the mental model) and eventually I realize the piles belong
together. So I merge them.

---

[1] Wikipedia definition: "The universe of the game of Life is an infinite two-dimensional grid of
cells, each of which is either alive or dead. Cells interact with their eight neighbours, which are
the cells that are directly horizontally, vertically, or diagonally adjacent." (tinyurl.com/4ubdr)

[2] tinyurl.com/2bt5ke

Can I Hurry It Up?

Q: Can I finish one part of the diagram before the others?

A: No, it's like being pregnant. You can't give birth to an arm first ... okay, maybe I shouldn't use that analogy. Since the patterns keep shifting as new tasks are mixed in, you don't have a fully baked diagram until you have finished incorporating all the tasks.

To illustrate what I mean about shifting patterns, I have a series of snapshots of the moviegoer mental model. In these images, you can see a certain set of tasks and groups morphing over time as I have added data. I was using a word processing document to track the tasks. I put placeholder headers in the document to act as parking lots for tasks I knew *weren't* in the groups I had already defined. These parking lot headers were "Tower" and "Mental Space." There was also a parking lot of orphan tasks, called "Task—Still Needs Processing!" (See Figure 9.4.)

Mental Space
- **Tower**
  - **Discuss Film Afterwards**
  - **Let the Movie Linger**
  - **Write a Review**
  - **Track Production Studio/Box Office News**
  - **Get Toys, DVDs, Books, Soundtracks**
  - **Wish that a Film Could Change How You Act**
  - **Investigate Story from a Film Afterwards**
- **Tower**
  - **Have a Date Night**
  - **Watch Films Regularly**
  - **Organize a Group**
  - **Attend Films with Others**
  - **Attend Films Alone**
  - **Read the Book**
  - **Listen to Soundtrack**
  - **Study Film**
  - **Write Entertainment**
  - **Task – Still Needs Processing!**

FIGURE 9.4.
Parking lot headers "Tower" and "Mental Space."

If I added a task to a parking lot where there was another similar task, I pulled the two out and started a new group. You can see the same document a few days later (Figure 9.5), after I had added the tasks from a few more transcripts. Note that some of the same tasks exist, but have been re-arranged in different groups, and some of the towers have been re-named. A new mental space has appeared named "Get Reviewers' Opinion of Films."

Mental Space

Tower
Discuss Film Afterwards
Let the Movie Linger
Write a Review
Track Production Studio/Box Office News
Get Toys, DVDs, Books, Soundtracks
Wish that a Film Could Change How You Act
Investigate Story from a Film Afterwards

Dress in Costume
Dress in Costume

Tower
Have a Date Night
Watch Films Regularly
Organize a Group
Attend Films with Others
Read the Book
Listen to Soundtrack
Study Film
Write Entertainment

Watch Film Alone
Watch Certain DVDs Alone
Attend Films Alone

Get Reviewers' Opinion of Films

Get Reviewers' Opinion of a Film
Read the Reviewers You Respect
Read Compiled Reviews
Read Many Reviews to Get a Consensus
Read Newspaper Reviews
Read Film Festival Reviews

FIGURE 9.5.
A few days and tasks later...

Next, in Figure 9.6, you can see something closer to the finished product. "Get Reviewers' Opinion of Films" has expanded to include other ways people learn about a film. Hence, that mental space was renamed "Learn More About a Film." Other towers such as "Have a Date Night" have moved to new groups that formed. Note that Figure 9.6 does not show the tail end of the mental model, so tasks such as "Discuss Film Afterwards" did not disappear. They merely appear further down, out of range for this image.

Formatting Towers and Mental Spaces

So, how do you keep track of all these piles? There get to be a lot of them. I have had an average of 21 mental spaces in the models I've created in the past five years. If, on average, each mental space has 5 towers and each tower has 5 tasks, that's 525 tasks to put into piles. An average transcript from an hour-long interview contains between 60 and 120 tasks. Brace yourself! You'll have to keep track of where each of those tasks went.

If you are using sticky notes, a couple of conference tables pulled together provide enough workspace, or you can use a whole wall. I have seen people use big pieces of poster board so their work is semi-portable. You can cluster sticky notes on top of each other, which saves space. You'll just have to un-stick them when you want to examine whether a particular pile needs to be pulled apart or merged with another pile. You can use different colored sticky notes to indicate the title of a pile (a tower) and the name of a set of piles (a mental space). The sticky-note approach to grouping is particularly good for co-located teams who want to spend a week or two in the same room building the model. Of course, co-located teams also can use the electronic approach with great success.

Decide to Watch a Film
Watch a Film for Pleasure
Watch a Film to Avoid Something
Watch a Film to Achieve a Mood
Have a Date Night
Take Family Dependents to a Film
Share a Film with Friends
Watch a Film to Learn
Watch a Film in Theater for the Experience
Watch Films Regularly
Watch a Film Spontaneously
Attend Films with a Group
Watch Your Film Expenses

Encounter a Film You Haven't Heard Of
Encounter a Film You Haven't Heard Of

Choose Films
Look for Certain Qualities
Choose Films because of Familiarity
Watch a Certain Genre During Winter
Include Random Film Choices
Inherit Interest in a Genre from Family
Choose Films Together with Companions
Avoid Certain Films
Decide on Film Based on Reviewer's Opinion
Postpone a Film

Learn More about a Film
Read Review to Get an Idea of the Film
Talk About Films You Want to See
Look for Trailers Online

FIGURE 9.6.
Part of the nearly finished mental model.

Here are three examples of how practitioners have used sticky notes to look for patterns and group concepts. In Figure 9.7, the practitioners arranged sticky notes on tall, portable poster boards. The format uses vertical towers of tasks just as in the mental model diagram. Tower names were assigned with blue sticky notes.

FIGURE 9.7.
Photo courtesy of Mary Piontkowski while at Macy's.

In Figure 9.8, the practitioners literally cut quotes from a Word document with a pair of scissors and piled the slips of paper by affinity. The practitioners found it easy to move tasks from one pile to another quickly. Once they finished sorting the quotes, they labeled them as tasks with small sticky notes, then took each pile and looked for subdivisions to make towers or looked for similarities to merge piles. This research was gathered and analyzed in French in Montréal.

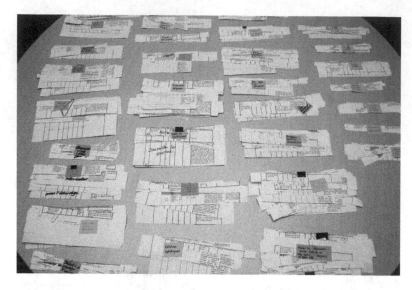

FIGURE 9.8.
Photo courtesy of Isabelle Peyrichoux and Vanessa Joanes, User Experience Group, Bell Web Solutions.

In Figure 9.9, Sarah Landelle of the United Nations writes potential tower names on erasable sheets of electrostatic plastic[3] on the wall. Sticky notes are grouped on the sheets.

If you are taking the electronic approach, you will want to use the outline feature of a word processor or spreadsheet. An outline allows you to collapse many lines in a section into a single line. You can later expand that single line into the whole set by clicking an icon in the left margin. Collapsing data like this allows you to see the whole structure that is forming without too much scrolling. It makes it easier to answer the question, "Where did I *put* those tasks like this? I know I have them in here somewhere!"

In your electronic document, you will want to have columns (reading from the right) for the quote, the participant data, and the atomic task. If you are using a spreadsheet, you also will want a column for the task, for the tower,

---

3 Purchase the sheets at tinyurl.com/2d6uuw and read more about a similar product in a blog entry by Kate Rutter of Adaptive Path: tinyurl.com/2ffftg

FIGURE 9.9.
Photo courtesy of Craig Duncan, United Nations.

and for the mental space. If you are using a word processor, you can use heading level three for the task names, heading level two for the towers, and heading level one for the mental spaces. In a word processor, outlining depends upon having headings. If there are no headings, then your lines won't collapse into anything. In a spreadsheet, you can collapse any set of rows. No headings are necessary in a spreadsheet.

In a word processor, there is usually a menu item for selecting how you want to view the document. In the 2003 version of Microsoft Word, this menu is called the View menu, and it contains selections like Normal, Print Layout, and Outline. To view documents in Outline view properly, every single line in the document needs to be assigned either a heading style or a normal paragraph or table style. This is the way in which the word processor keeps track of the hierarchy of information to display.[4] In the following example, Figure 9.10, you see a document in Outline view with three levels of headings displayed.

---

4   For more information about creating and applying heading styles, see the Microsoft site.

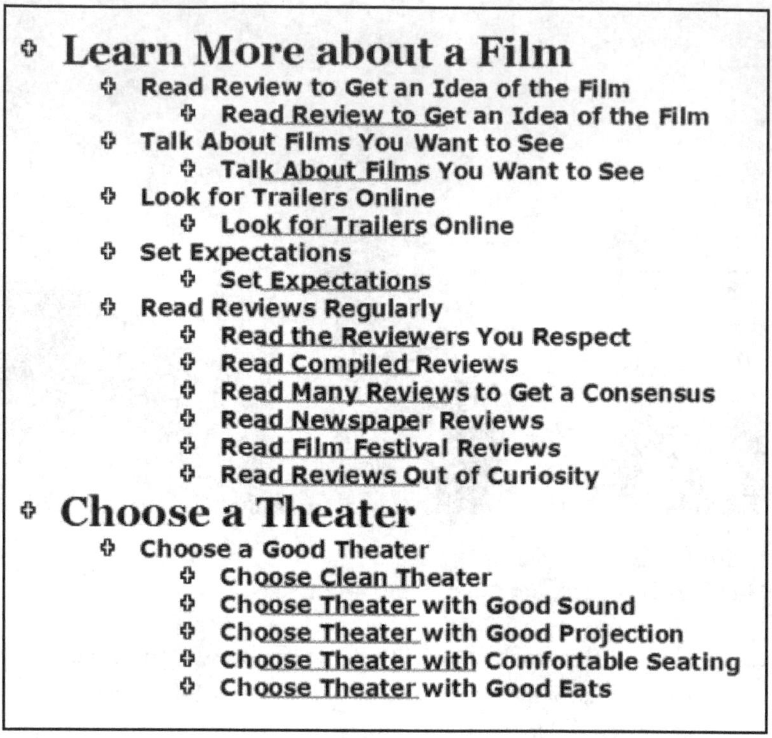

```
⊕ Learn More about a Film
    ⊕  Read Review to Get an Idea of the Film
        ⊕  Read Review to Get an Idea of the Film
    ⊕  Talk About Films You Want to See
        ⊕  Talk About Films You Want to See
    ⊕  Look for Trailers Online
        ⊕  Look for Trailers Online
    ⊕  Set Expectations
        ⊕  Set Expectations
    ⊕  Read Reviews Regularly
        ⊕  Read the Reviewers You Respect
        ⊕  Read Compiled Reviews
        ⊕  Read Many Reviews to Get a Consensus
        ⊕  Read Newspaper Reviews
        ⊕  Read Film Festival Reviews
        ⊕  Read Reviews Out of Curiosity
⊕ Choose a Theater
    ⊕  Choose a Good Theater
        ⊕  Choose Clean Theater
        ⊕  Choose Theater with Good Sound
        ⊕  Choose Theater with Good Projection
        ⊕  Choose Theater with Comfortable Seating
        ⊕  Choose Theater with Good Eats
```

FIGURE 9.10.
Microsoft Word document with three levels of headings in Outline view.

If you double-click on the plus icons to the left of the task "Set Expectations" (not the tower), you will expand that section to see the task table, as shown in Figure 9.11.

You can expand more than one area at a time (see Figure 9.12).

In Word, you double-click the plus icon once again to collapse the section. There is a set of controls that usually appear in the Word toolbar when you are using Outline view (Figure 9.13). These include a series of icons for increasing or decreasing the heading level of a selected line, as well as a droplist that contains the phrases "Show Level 1," "Show Level 2," through "Show All Levels." It is the droplist that you will use; you can safely ignore the other controls.

✧ **Learn More about a Film**
    ✧ **Read Review to Get an Idea of the Film|**
        ✧ **Read Review to Get an Idea of the Film_____**
    ✧ **Talk About Films You Want to See**
        ✧ **Talk About Films I Want to See**
    ✧ **Look for Trailers Online**
        ✧ **Look for Trailers Online**
    ✧ **Set Expectations**
        ✧ **Set Expectations**

Atomic Tasks	A	R	Supporting Quotes
Set Expectations Low to Avoid Being Disappointed	C	S F	Going to see movies is very much about setting expectations ahead of time before you see it... I don't enjoy being constantly disappointed
Set No Expectations	F P	S R	I might have expectations and be disappointed
Know What Kind of Film You're Spending Money On	F P	S F	you ought to know what you're spending money on

FIGURE 9.11.
Outline view with a section expanded to see the task table.

The examples I show in Figures 9.11 and 9.12 are set to Show Level 3. This setting lets me scroll quickly through the structure of the document. All the tasks fit within four pages instead of 45 pages.

Figure 9.13 shows what a spreadsheet in Outline view looks like, with the same two sections expanded as in the Word example in Figure 9.12.

Here you see that the task, tower, and mental space each have columns, instead of being represented by headings. In the left margin, a series of plus and minus signs appear, along with long vertical bars. The plus sign signifies that there are rows collapsed and hidden beneath this row.

Clicking the plus sign expands those rows. The minus sign and vertical bar signify that the rows encompassed by the vertical bar are expanded and belong to the line with the minus sign on it. If you click the minus sign, those rows will collapse and hide, and the minus will change to a plus.

⇧ **Learn More about a Film**
   ⇧ **Read Review to Get an Idea of the Film**
      ⇧ **Read Review to Get an Idea of the Film**_____
   ⇧ **Talk About Films You Want to See**
      ⇧ **Talk About Films I Want to See**
   ⇧ **Look for Trailers Online**
      ⇧ **Look for Trailers Online**
   ⇧ **Set Expectations**
      ⇧ **Set Expectations**

Atomic Tasks	A	R	Supporting Quotes
Set Expectations Low to Avoid Being Disappointed	C	S F	Going to see movies is very much about setting expectations ahead of time before you see it... I don't enjoy being constantly disappointed
Set No Expectations	F P	S R	I might have expectations and be disappointed
Know What Kind of Film You're Spending Money On	F P	S F	you ought to know what you're spending money on

   ⇧ **Read Reviews Regularly**
      ⇧ **Read the Reviewers You Respect**
      ⇧ **Read Compiled Reviews**

Atomic Tasks	A	R	Supporting Quotes
Read Compiled Reviews	F P	S F	go look at a website called Metacritic which compiles movie reviews... Rather than reading reviews in local papers, we cast a broader net
Avoid Reading Individual Reviews	F P	S F	don't read the individual reviews
	F P	S R	if the critical consensus is really high, I'll forgo reading the review

FIGURE 9.12.
Two task tables expanded at the same time.

Collapsing to the row below a selected group of rows is the default for most spreadsheets, since it assumes you are calculating a sum of numbers and only want to display the total at the bottom of the column. You will have to change this to collapse a group to the row above. In the 2003 version of Microsoft Excel, you do this by selecting "Settings" from the Group and Outline choice in the Data menu. In the window that appears, simply un-check the two settings. The first setting, "Summary rows below detail," is the *totals* row into which a group of rows will collapse in the outline. Un-check the second setting as well, "Summary columns to right of detail," although it does not really make a difference for your purposes.

I don't ordinarily group the atomic tasks under each task, nor do I always group the tasks under each tower. It is often sufficient to group all the rows under each mental space. When there are a lot of towers, I will group the Tasks under them to keep from scrolling too much. So, the fact that there are three levels of plusses in Figure 9.14 is rare. Usually there are one or two levels. (Now that I've said all this about outlines in Excel, I have to admit that sometimes I skip them entirely. Sometimes it's less awkward to just scroll or use search to find what I want.)

Quick Answers about the Examples

The AS in the table stands for Audience Segment. The R stands for Region. You might have noticed, too, that the first four towers (heading level two) have the same name as the task (heading level three). You want to propagate the vocabulary from the lowest levels up to the highest levels. If there are several tasks in one tower, I make up a title that incorporates words from all the tasks. If there is only one task in the tower, I usually copy the title exactly.

Decide If You Need More

By now you should have a good feel for your data. You know what patterns keep repeating themselves. The data has grouped itself into towers and mental spaces. But what happens if you have some leftover tasks that just don't go with anything else? You've made sure these extraneous tasks are root tasks, but there just doesn't seem to be any other data that matches each of them. If the topic is important and within your scope of work, this set of single-task towers is an indication that you might want to conduct another interview or two in the topic area.

1	Mental Space	Task Tower	Task	Atomic Task
2	Decide to Watch a Film			
92	Encounter a Film I Have Not Heard Of			
105	Choose Films			
201	Learn More about a Film			
202			Read Review to Get an Idea of the Film	
205			TalkAbout Films I Want to See	
208			Look for Trailers Online	
211			Set Expectations	
212			Set Expectations	
213				Set Expectations Low to Avoid Being
214				Set No Expectations
215				Know What Kind of Film I Am Spending Money On
216			Read Reviews Regularly	
217				Read the Reviewers I Respect
219				Read Compiled Reviews
220				Read Compiled
221				Avoid Reading Individual Reviews
222				
223				Read Many Reviews to Get a Co

FIGURE 9.13.
Microsoft Excel example of outline view.

Aud Seg	Loc	Quote
C	SF	Going to see movies is very much about setting expectations ahead of time before you see it... I don't enjoy being constantly disappointed
FP	SR	I might have expectations and be disappointed
FP	SF	you ought to know what you're spending money on
FP	SF	go look at a website called Metacritic which compiles movie reviews... Rather than reading reviews in local
FP	SF	don't read the individual reviews
FP	SR	if the critical consensus is really high, I'll forgo reading the review
onsensus		

In general, once your patterns have settled down, you can be reassured you have captured all of the mental spaces. Any amount of additional interviewing will not add another mental space. Additional interviews might add more tasks to existing towers, and there is a chance you'll discover a new tower or two, but you can rely upon the mental spaces that have been defined.

Database of Tasks

I was contacted by a team in 2003 that had put all the tasks they had combed into a database. I thought that was great because I am tool agnostic: use a text document, a spreadsheet, sticky notes, a database...sure! But then I realized the reason they had entered all the tasks in a database was to automate the grouping process. Their first pass was to alphabetize the entries by verb. This attempt at grouping failed because there were different verbs that meant the same thing. The alphabetical order of the verbs had no bearing on the relationship between the entries, nor did sets of the same verb represent only one concept. I advised the team that grouping can't be automated. You have to think hard about each task and look for affinities to other tasks.

Later, another group, lead by my collaborator Mary Piontkowski, had a little more success. While combing the first few transcripts, they assigned a tag to each atomic task. These tags were things like "decide," "plan," "research," and "compare." The team extrapolated upon the tags, "plan x," "plan y," "plan z" and made up an official list that they used to tag all the rest of the atomic tasks they combed out of subsequent transcripts. If an atomic task didn't match one of the tags, they added a new tag to the list and let everyone know. After combing, the atomic tasks were already somewhat sorted into conceptual groups. They still had to step through each task, assess its meaning, and assign it to a group, but they were able to split the work among three people because they could assign related tag sets to each person. It made grouping feel less complicated.

Plan Your Logistics

You *can* create a mental model by yourself, but sometimes it takes a village. There are different roles that team members can play, and I have a few tips on how to make the process run smoothly.

There are three main roles: interviewer, comber, and grouper. (These sound like types of fish you want to avoid ordering at a restaurant...) Depending on time available, schedule, and personality, one person can play any number of the roles. For example, Anne and Barbara can conduct interviews, Eric can start combing Anne's transcripts as they become available, and after she's finished with interviews, Anne can comb Barbara's transcripts. Meanwhile, after completing the interviews, Barbara can get started grouping the information already combed out of the transcripts by Eric. Here's a sketch of how their tasks would fall in relation to one another (Figure 9.14).

Logistics Examples				
Task	**Week 1**	**Week 2**	**Week 3**	**Week 4**
Anne				
Conduct Interviews	▮			
Comb Transcripts		▮	▮	
Group Concepts				▮
Barbara				
Conduct Interviews	▮			
Group Concepts		▮	▮	▮
Eric				
Comb Transcripts	▮	▮		

FIGURE 9.14.
Project plan for two interviewers, two combers, and one grouper.

You can divide up the work in many ways. You could have the same two people tackle each task, like Carol and Fran in Figure 9.15. You could have one person interview and group, while a second person combs all the transcripts, like Kate and Gina in Figure 9.15. In another scenario, David and Lisa do all the interviewing and combing, but Lisa completes the grouping alone. Or you could do the opposite, where Heather does all the interviewing alone, but has help from Amanda for combing and grouping. You could have one person, Mimi, in charge of interviewing and grouping, while others do the combing, as in the sixth row of Figure 9.15. You could have a pair doing interviews, Aaron and Catherine, one of those people plus a new member combing, Catherine and Donna, and all three grouping. Or, more difficult to pull off from a continuity perspective, you could have different people involved in each stage, as in the last row of Figure 9.15.

Often one person is in charge of the grouping, since you must keep track of where you've put every task. If multiple people do the grouping, each must still keep track of every task. "Divide and conquer" among groupers does not work well. If you do divide sets of tasks and assign them to different people for grouping, you will lose the chance to combine tasks among different individuals' piles. Unless you have a telepathy server installed, it will become much more difficult to let the patterns morph and grow naturally. Putting one person in charge of reviewing all the groups being created by others, and pulling things out of their groups when necessary, alleviates this problem.

INTERVIEWER	COMBER	GROUPER
Anne Barbara	Anne Eric	Barbara
Carol Fran	Carol Fran	Carol Fran
Kate	Gina	Kate
David Lisa	David Lisa	Lisa
Heather	Amanda Heather	Amanda Heather
Mimi	Ingrid Jay Charles	Mimi
Aaron Catherine	Catherine Donna	Aaron Catherine Donna
Amy Bob	Carrie Denise Eleanor	Francisco Glenn

FIGURE 9.15.
Role combination examples.

Another possible workaround is to try dividing the day into parts where each person completely owns the document during her part of the day, then passes it on to the next person. I have seen this work reasonably well when each person highlights their new entries with a certain color and colors big shifts in the patterns with another color. This way the next person can easily see what has changed in the document. A little extra time should be allowed for each person to get familiar with the changes from the previous person.

Comb Early and Often

If possible, begin combing some transcripts before you complete all
the interviews. I say this mostly because it helps maintain momentum.
The interviewing might exhaust you, and if you also play the role of a
comber, you will be tempted to take an ever-lengthening break between
interviewing and combing.

It's not a good idea to start grouping, however, until you have conducted
all the interviews. You run the risk of letting the structures and patterns
you see in grouping influence the way to ask questions in an interview. Of
course, if you are in the rare situation where the people grouping are not
conducting interviews, then this warning does not apply.

There are several ways to approach combing. The first three that I describe
assume that combers are working individually. The simplest approach
is to give one transcript to each comber and ask each to pull the tasks
into a separate document. When finished, each comber then hands their
document over to the person in charge of the grouping. When you give
a comber a new transcript, she then pulls the tasks into another new
document. If you conducted 24 interviews, there will be 24 transcripts and
24 separate task documents that the grouper will pull together.

Another approach, which is more instructive, allows the combers to comb
all of their transcripts into one document per person. This allows a comber
to combine new tasks with the tasks she has previously recorded from
other transcripts.

An even more collaborative way is to give each comber a version of the
master document that the grouper is creating to comb tasks into. Before
beginning to comb a new transcript, the comber asks for a copy of the
grouper's document, and combs into that. This approach requires that
the grouper be a separate person from the combers, and also requires
that the grouper be working in parallel.

Instructions to Individual Combers

When combing into a document:

1. Everyone combing a transcript uses a different font or highlight color. You can choose whatever color you want, so long as it's unique to you, and you can read the text easily.

2. Go through the transcript line by line and highlight phrases that you think represent a task. Change the color of these highlighted phrases to "your" color.

3. For each phrase you highlight and color, copy it to a document where its color will be preserved. In the document, label it with the transcript number, the audience segment, and the location. Then, if possible, label the phrase with a verb+noun phrase to describe the task. Try to use words from the quote.

When combing with sticky notes:

1. Everyone combing a transcript uses a different color pen.

2. Go through the transcript line by line and circle phrases with your colored pen that you think represent a task.

3. For each phrase you circle, write a verb+noun phrase that describes the task on a sticky note. Annotate the sticky note with the transcript number, the audience segment, and the location. Try to use words from the phrase you circled.

Taking It All In

If your interviewers are also your combers, I have experimented with swapping transcripts with other interviewers. We each comb an interview we did not conduct. This swap allows us to read what happened in the other interviews.

If your combers want to work as a group, there are two approaches to combing. The first approach is simple: Sit down together with one transcript, read it aloud, agree on the tasks you find, and record them in a document. Get the next transcript, read it out loud, and add these tasks to a different new document. The second approach is a variation on the first, where the group adds tasks to the same document, noting where tasks are the same and placing them next to each other. After all the transcripts are combed, the grouper(s) then take over. If the combers are also the groupers, then they just move into the next stage.

Note that in all the above situations, you can replace the word "document" with "sticky note." If all that writing by hand alarms you, you can print to sticky notes. Recently, 3M has put their Post-it™ notes on a page that can fit in a printer. Packages are sold in sheets of 10, with six notes per page.[5] If you search for "3M Printscape Personalized Note Kit" you should find some sources to purchase from.

---

5 Mary Piontkowski's Word template is available under Resources on the book site. It will help you format what you want to print into 3x4-inch rectangles. 🐘 www.rosenfeldmedia.com/books/mental-models/content/resources

All At Once

Years ago, I used to comb for tasks and group all at the same time. This is feasible when you are the only one doing the whole process. Yet, it was inexplicably overwhelming. Among other frustrations, it was hard to feel a sense of accomplishment, of moving towards the goal. By breaking the process into two steps, combing and grouping, I was able to see progress more clearly; moreover, I was able to make better estimates of how long the process would take.

You will find details about how long it might take to complete combing and grouping in Appendix A, "How Much Time and Money?" available under Resources on the book site at www.rosenfeldmedia.com/books/mental-models/content/resources

Congratulate Yourself

The mental model combing and grouping process can feel overwhelming when you're in the middle of it. I always feel like it will go on forever, and at some point I start to panic. You'll probably feel these emotions, too. But it's a finite set of data, and you will eventually get through it. Once you're through, you can breathe a sigh of relief. The hardest work is behind you. Next, you will render your tasks, groups, and mental spaces in diagram format.

CHAPTER 10

Create the
Mental Model

Build the Model Automatically   198
Build the Model Block-by-Block   200
Review the Diagram with Project Guides   216
What Did You Learn?   218
Decorate the Diagram   218
Ask for Feedback   223

After combing and grouping, building the mental model is almost a relief. The structure for the diagram already exists in your grouped document. All you need to do is render it as a diagram, and then spend time with stakeholders and other team members reviewing and polishing it.

The format of the diagram is malleable. In fact, the data behind the diagram can be grouped together in different ways to represent other concepts, such as workflow. The main point of the horizontal format that I use is to match features and content to each tower easily. I purposely don't use the horizontal format to represent cycles or step-by-step chronological order processes. Neither of these is necessary when your only goal is to map out your product beneath the mental model. Overloading the diagram with this information would be a mistake. Feel free to use the same data in other visual formats to represent workflow or other concepts.

After you and your team are happy with the diagram, I encourage you to decorate it to a certain extent. Information design comes into play when choosing what to represent in the diagram—that is, certain tasks might be country-specific, or perhaps only one audience segment performs a certain sub-section of the diagram. You may wish to show a division between female and male or some such task-related division. I will show a few examples in this chapter, and then refer you to Tufte[1] in order to free your creative spirit.

Build the Model Automatically

It's true. If you did your grouping correctly, you already have your mental model. I have a Python[2] script that interprets either a Microsoft Word or Excel document into an XML file that either Omnigraffle or Microsoft Visio can open. (Technically, the script creates a VDX file, which is Visio's version of XML. Omnigraffle can open this file too.) This script helps in two ways: One, it makes creating the diagram a matter of a few minutes.

---

[1] Edward R. Tufte gives enormously popular and inspiring seminars on information design and has written several books about the topic. Find out more at www.edwardtufte.com

[2] Python is a scripting language much like Perl. You can download the latest binary release for your platform from tinyurl.com/lo4uk

Two, it allows you to fiddle with the data in the original Word or Excel format again and again, then generate a new diagram each time with the press of a button. The days when I had to shift whole mental spaces aside in the diagram to make room for a new tower are gone. Believe me, that took patience. Diagramming applications don't scroll very fast when you have a large set of shapes selected.

I am releasing this Python script on my book site with the following caveat: It is pretty finicky about the format of the Word or Excel file, and I won't be able to help you debug why it won't slurp up your file and spit out a nice mental model diagram. For example, it doesn't know what to do with a space that comes at the end of a heading in Word, so it gets frazzled. And, if your column headings in Excel aren't exactly right, it will poop out before processing anything. The screen will flash and you will be left with...nothing.

Instructions for installing and running the script, as well as templates formatting the data that the script expects to see, are in the Resources section of my book site. ₥ www.rosenfeldmedia.com/books/mental-models/content/resources

In case you can't use the script, here is a quick lesson in how to create the diagram by hand. Or if you want to sketch one by hand, these instructions will help.

Automatic Coloring Using the Script

There are a few ways for Microsoft Excel to define the style of a cell. It can do it in any of the tags: ‹Table›, ‹Column›, ‹Row›, ‹Cell›. At the moment, the script only checks the style defined in the ‹Cell› tags. Hence, if you set the background color of a cell, it will affect what color that tower or task is in the resulting diagram. The script does not check for text color; all the text appears in black font.

Moreover, if you color the entire row, both the task and the tower will inherit that color. If you want different colors for towers, you will need to color those cells separately.

It is possible that the script has evolved since the printing of this book. Check my book site for updates: ⚑ **www.rosenfeldmedia.com/ books/mental-models/content/resources**

Build the Model Block-by-Block

Using a drawing program such as Omnigraffle or Microsoft Visio, start by creating small rectangles. Make sure that the rectangles are sized to have the best chance of providing enough room for your task titles. I use rectangles that are 5/8 of an inch wide (0.625 inch, 1.5875 cm) and 3/8 of an inch tall (0.375 inch, 0.9525 cm). Using a 6-point Arial font, this size box can contain four lines of text.

Fitting Text in Those Little Boxes

Sometimes your task title does not fit in four lines of 6-point font. You are already trying your best to be clear-yet-terse with the task titles. What can you do in case your text wraps to five lines? Try abbreviations, using "w/" instead of "with," or deleting articles such as "the." If that doesn't do it, use common industry abbreviations or acronyms that the audience of your diagram is likely to understand. Or substitute shorter words, such as "uneasy" for "uncomfortable" if it doesn't veer from actual vocabulary.

Make a few of these boxes and fill them in with the titles from the task level (third level) of your document (see Figures 10.1 and 10.2). Be sure you are pulling the tasks, not the atomic tasks.

Identify with a Film

Allow a Film to Permeate Your Life
- Buy the Soundtrack
- Listen to Soundtrack
- Read the Book Afterward
- Investigate Story from a Film Afterward
- Wish that a Film Can Change How I Act
- Let the Movie Linger

Collect Film-Related Stuff
- Save Tickets in Scrapbook
- Collect Film Artwork
- Collect Toys

Get the DVD
- Buy Special DVDs
- Get DVDs as Gifts

Watch a Film Multiple Times
- Watch DVDs You Own More than Once
- Watch Film Multiple Times

Interact with People about Film

Discuss the Film

FIGURE 10.1.
Microsoft Word format document with tasks in red.

Mental Space	Task Tower	Task	Atomic Task	Aud Seg	Loc
Identify with a Film					
	Allow a Film to Permeate My Life				
		Buy Soundtrack			
			Buy the Soundtrack	SM	SF
				FP	SR
				FP	SR
		Listen to Soundtrack			
			Listen to Soundtrack	SM	SF
				FP	SR
		Read the Book Afterward			
			Read Book After Seeing the	FP	SR
			Buy the Book	FP	SR
		Investigate Story from a Film Afterward			
			Investigate Story from a Film Afterwards Out of Curiosity	FP	SR
		Wish that a Film Can Change How I Act			
		Let the Movie Linger			
	Collect Film-Related Stuff				
		Save Tickets in Scrapbook			
		Collect Film Artwork			
		Collect Toys			
	Get the DVD				
		Buy Special DVDs			
		Get DVDs as Gifts			
	Watch a Film Multiple Times				
		Watch DVDs I Own More than Once			
		Watch Film Multiple Times			
Interact with People about Film					
Follow the Industry					

FIGURE 10.2.
Microsoft Excel format document with tasks in red.

Stack the boxes that are all related to one another by group. Then set the boxes so that each left edge is lined up with each inch mark on your grid (see Figure 10.3). Work with sets of four or five towers, or whatever will fit on your screen without scrolling. You can either place all the tasks first, or work left to right placing tasks, towers, and mental spaces.

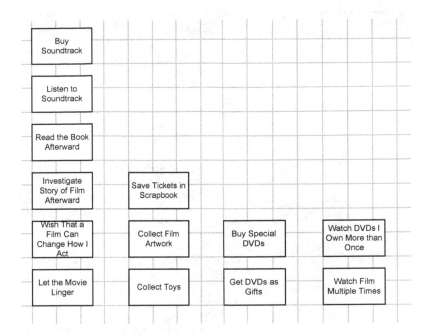

FIGURE 10.3.
Stacked task boxes aligned to each inch in the grid in Omnigraffle or Microsoft Visio.

Next, create the towers. The towers I use are 7/8 of an inch wide (0.0875 inch, 2.223 cm) and as high as they need to be to include all of the task boxes plus the tower title. Put the tower box behind the task boxes by using a "Send to Back" command. If there are so many tasks that the tower will, ahem, tower above the others, then I make a double-wide tower and place the tasks in two columns within it. Generally speaking, I allow my towers to be seven tasks high. I often give the tower a default background color and remove the boundary line. Some people give the boxes rounded corners. You can do as you see fit for now. Later I will discuss assigning colors that represent audience segments or regions or other significant data.

Mental Space	Task Tower	Task	Atomic Task	Aud Seg	Loc
Identify with a Film					
	Allow a Film to Permeate My Life				
		Buy Soundtrack			
		Listen to Soundtrack			
		Read the Book Afterward			
		Investigate Story from a Film Afterward			
		Wish that a Film Can Change How I Act			
		Let the Movie Linger			
	Collect Film-Related Stuff				
		Save Tickets in Scrapbook			
		Collect Film Artwork			
		Collect Toys			
	Get the DVD				
		Buy Special DVDs			
		Get DVDs as Gifts			
	Watch a Film Multiple Times				
		Watch DVDs I Own More than Once			
		Watch Film Multiple Times			
Interact with People about Film					
	Discuss the Film Afterward				
		Avoid Discussions			
		Ask Strangers Their Opinion After a Film			
		Discuss Interpretation of Book			
		Learn Craft from Discussion of Unusual Point			

FIGURE 10.4.

Microsoft Excel format document with towers in red.

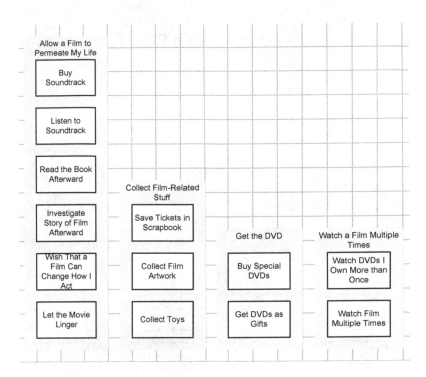

FIGURE 10.5.
Tasks with towers.

Next, draw thick vertical lines that bracket this mental space, and add a title roughly centered above the towers (as shown in Figure 10.6).

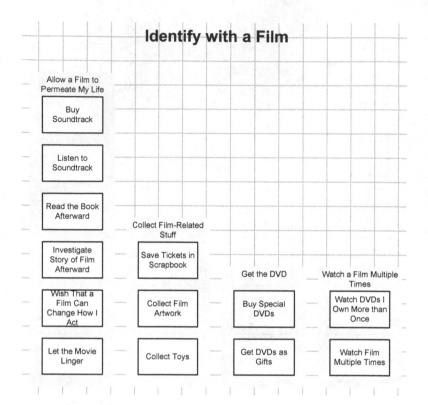

FIGURE 10.6.
Tasks in towers in a mental space.

After you get all of the towers and mental spaces in place, draw a thick horizontal line beneath all the towers (as shown in Figure 10.7). This line will divide the upper mental model from the boxes you will align beneath it later.

Finally, add a legend to the diagram. The legend that I use addresses the placement of items in the diagram, as well as colors used. I usually augment the legend after I have aligned content and features underneath the towers. For now, the legend I begin with looks similar to that shown in Figure 10.8.

I don't always represent the number of voices per task. Because the research is qualitative, representing the popularity of a task could be misleading. You don't ask the same questions of every interview participant; therefore, you might not have an answer from someone whose voice might count towards a certain task. And, if you interview another 44 participants, the tasks that are popular might change. Nonetheless, I find it interesting to indicate which tasks I've heard a lot about. The rare task, if I depict it, is always the single voice—that is, just one person mentioned it. The numeric threshold between a "normal" and a "popular" task varies depending on how many people you interview. I usually select a threshold based on a desire to point out a few interesting trends that I notice, making sure that "popular" tasks are as rare as the single-voice tasks.

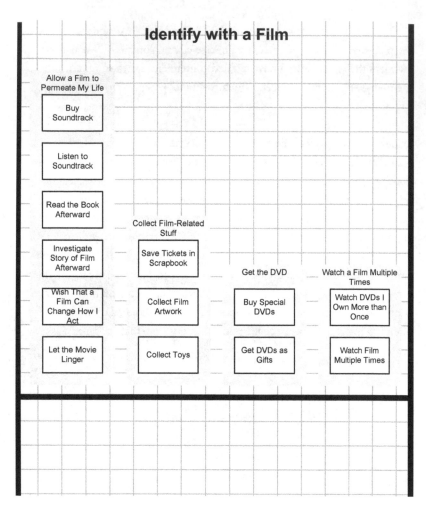

FIGURE 10.7.
Part of a finished mental model. See the finished moviegoer mental model on the book site: 🐘 www.rosenfeldmedia.com/books/mental-models/blog/moviegoer_alignment_diagram

Interact with People About Film

Discuss the Film Afterward

- Avoid Discussions

- Ask Strangers Their Opinion After a Film

- Discuss Interpretation of Book

- Learn Craft from Discussion of Unusual Points

- Discuss Film Afterward

- Go Somewhere to Sit and Discuss Film

Recommend a Film

- Recommend a Film when Asked

- Write a Review

Follow the Indu

Track Box Office Competition

- Track Box Office Competition

Track Production Studio News

- Track Production Studio News

Becom About

- V Enter

- Stu (

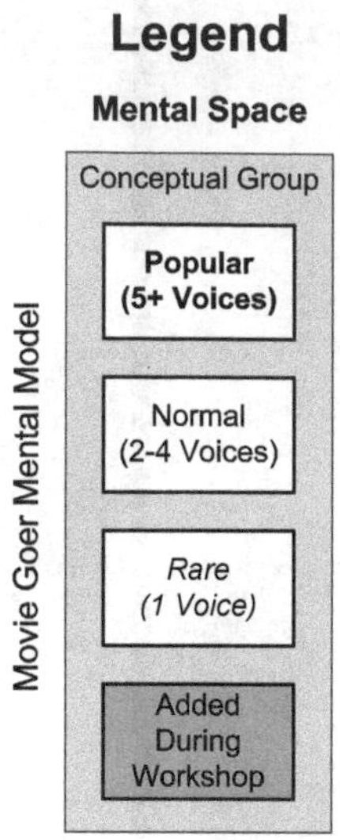

Legend

Mental Space

FIGURE 10.8.
Example legend.

Split the Model into Multiple Diagrams

Every organization needs to interact with several audiences, from employees and investors to customers and the general public. Rather than using a single solution, try to cater to audiences with different behaviors—every organization should create the right number of solutions that each makes sense to the audiences concerned. Each full mental model represents a separate entity. You can use the mental model to define the beginning and end—the boundaries—of your design. If each mental model delineates the beginning and end of one solution, then this concept is simply the idea of several mental models playing in concert to make up the whole offering of an organization.

Mental models are not simple things. Often there is overlap between what two different audience segments do. For example, purchasing agents and researchers both order chemicals for use in R&D labs. Both of them look for the chemicals with the properties that will work for what the scientist wishes to achieve. However, the purchasing agent also looks for cheaper substitutes or bulk deals, whereas the scientist looks for different chemical properties that might work better. There is a slight difference, but most tasks are in common. Then again, a user might migrate from one set of tasks to the next, such as a "manager" that is also an "employee." In this example, the two mental models contain different tasks and should be kept separate, but a user might cross over from one mental model to the other frequently or within a short period of time. What do you do in the case of the overlapping tasks?

How to Read a Mental Model

Mental models are made up of three levels. If the diagram looks like a city skyline, mental spaces form the blocks, towers form the buildings, and tasks form the windows in the buildings. The diagram is built from tasks combed from a set of interview transcripts. While you build the diagram from the lowest level to the highest level, you read it in reverse.

Start by looking at the mental spaces and understanding the differences between them. Study each mental space separately, as each should stand on its own, not necessarily related to mental spaces to the left or right. There may be a slight chronological order to a few of the mental spaces, but largely they are listed in clumps that will make our next step easier: aligning existing and planned product features to the towers.

Next study the towers within each mental space and understand the differences between them by reading the task examples within them. Get a feel for where people focus when accomplishing the goals in this particular mental space. Review the audience, regional, and other differences between the towers. What stands out? Which towers are items you would not have thought of before the research? Does the whole mental space seem consistent with what you know of your customers' world?

And what do you do when there are hardly any tasks in common, such as the "truck dispatcher" and the "service center mechanic." An application can be designed for each of these audiences with the specific tasks of the user in mind, following a less-is-more principle. Users will completely grasp the utility of the application and will not be distracted or confused by features intended for a different audience. Gone are the days of the monolithic application that tries to do everything for everyone, making it hard for anyone to do anything simply. How many times has, "But I *just* want to ____!!!" been uttered to a computer screen in frustration?

My rough rule of thumb is that if there is "a lot in common," then I keep the items in the same diagram or merge two diagrams. If there is "little in common," then I separate the tasks into two (or more) diagrams. When I separate tasks into two mental models, I mark the tasks that are common to the other audience segment with color-coding of the tower or the tasks.

On the flip side, sometimes the diagrams are just too long to understand. For example, there are three roles involved in purchasing large enterprise software suites: the evaluator, the implementer, and the maintainer. Sometimes one person plays all three roles; sometimes there are a few people playing these roles, with project hand-off among them. In these situations, I divide the diagram based on where the project hand-offs occur. I color-code the towers in each diagram with the roles that performed them, so that readers could see that an evaluator sometimes works far into the implementer's tasks. Three shorter diagrams are easier to comprehend.

The case of the folks who tend to move from one mental model to another requires deeper investigation. In this scenario, consider the frequency of shifts between the models and the distinction and intricacy of the task sets. In the case of the "employee" and the "manager," I recommend

How to Split or Merge the Mental Model

When many tasks are in common, *merge* mental models. Example: "researcher" and "purchasing agent."

When most tasks are different, keep mental models *distinct*. Example: "truck dispatcher" and "service center mechanic."

When users tend to shift from one model to another and back, keep mental models *distinct, but note the frequency in shifts* the user makes from one model to another and the complexity of each set of tasks. They might benefit from one merged solution with special parts for one audience. Example: "employee" and "manager."

When the diagram is too long to understand easily, *break it into subsets where hand-offs occur* between audiences, and consider whether the solution should be split as well. Example: "evaluator" and "maintainer" of an enterprise software suite.

one solution because the manager, who has the superset of tasks, moves between "check my vacation balance" and "open a job requisition" with cognitive ease. She sees both tasks as "administration." In this one solution the extra features supporting "open a job requisition," etc., appear only to those logged in as managers.

However, if the complexity of each set of tasks is deep you will need to devote a separate solution to each model. It turns out that the cognitive process an "evaluator" goes through to select enterprise-level software is entirely different than what the "maintainer" does. The former works pretty much up until the roll-out of the software, which is where the latter takes over. Yes, sometimes one person performed both roles, but because the tasks were so different, we preserved the subsets we had split the mental model into.

How to Split or Merge the Mental Model

If the frequency of shifts is high and users perform tasks in either set, then consider designing *one solution*. Example: "employee" and "manager."

If there is a big distinction between the types of tasks being done and each task set is deep and complex, create *separate solutions*, no matter how frequent the shifts are. Example: "evaluator" and "maintainer."

Splitting or merging diagrams is important because translation from mental model to solution is so direct. Everything that hangs together in one mental model is exactly what should hang together in one solution. In terms of intranets, organizations went astray—and still do—in this regard. For example, they created monolithic intranets that tried to hold everything that anyone employed by the company could want. If they had mapped out mental models for all the intended audiences, they would have easily seen that one solution was not the right approach. They should have implemented a galaxy of internal web properties each intended for a specific audience, with a central hub for use in case someone needed to look at information in another department. Thankfully, many organizations are not so naïve about their web properties anymore. Moreover, the corporate URL now represents a landing zone to choose which web property you wish to go to, as in www. yamaha.com and shown in Figure 10.9. Much traffic to a specific property comes through direct links, rather than this corporate landing zone.

An easy way to think of this is to use a campus analogy. If your company or school is large, you have many buildings, often co-located on one property—the campus. Other buildings may exist in other locations or countries, as a part of the virtual campus. Each building houses a

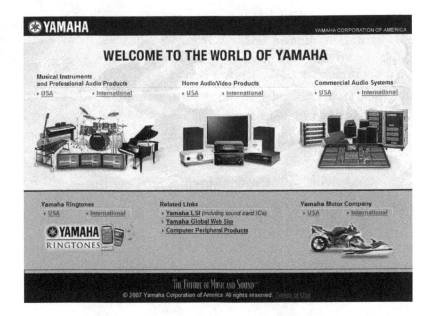

FIGURE 10.9.
The Yamaha landing page shuttles you to several web properties, including Musical Instruments, Home Audio/Video, Commercial Audio, Ringtones, and Motorcycles and Jet Skis.

particular department; everyone who works on a team is generally on the same floor. For your day-to-day business around the campus, you stick to one building, with a trip here or there to the cafeteria, or to engineering to meet with the development team. You are most familiar, however, with your own building. Most things you need for your daily work are there, organized in a way that more or less makes sense to you and your team.

Each one of your web properties is a building on your internet campus. Each property has its own unique navigation that represents the mental model of the people populating it. There is a utility navigation item that will take a person to a "shuttle station" or hub listing all the properties she can choose from as her destination.

Review the Diagram with Project Guides

After creating the diagram, review it in detail with all the stakeholders involved. The goal of this process is to give each person a feeling of ownership: "This is the mental model we made." If each person feels ownership for the diagram, then each person is likely to refer to it in discussions and decision-making. I also use this review workshop to pass along stories from the interviews—little anecdotes about a customer's daily life tend to be memorable.

Learn More about a Film

Legend

Mental Space

Movie Goer Mental Model

Conceptual Group
Popular (5+ Voices)
Normal (2-4 Voices)
Rare (1 Voice)
Added During Workshop

Read About Films I Want to See

Look for Trailers Online

| Look for Trailers Online |
| Look for Film Information Online |

Anticipate a Film

| Set Expectations |
| Get Excited about a Film |

FIGURE 10.10.
Diagram that has items that were added to the source document during a review workshop.

I run the review workshop with a projector and a remote connection for those not in the room with us. First, I scan through all the mental spaces, describing each of them in general. This high-level review usually takes 30 minutes. Then I go back to the left end of the diagram and step through all the towers, highlighting the most interesting tasks in each tower. In the beginning, I touch on every tower and we dive into explorations of what participants said. We debate the interpretation of certain tasks and make edits to the source document (word processor, spreadsheet, sticky notes). Sometimes we add a task that stakeholders have experienced but didn't get mentioned in the interviews with customers (Figure 10.10). For some of the tasks, we'll trace a quote back to the original transcript. During the first hour of the session, I make sure everyone knows the diagram is not cast in stone; people around the table typically have years of actual experiences with this user base that can be brought to bear. Towards the end of the detailed review, there is no need to read each tower and task aloud. At this point, I let the team read what's on the screen while I call out the more fascinating discoveries, unveil surprises, and recite interesting or funny stories. These stories are what make the workshop entertaining enough to endure, because the whole review usually takes between three and five hours.

These review hours are valuable when the stakeholders become co-creators of the diagram. If your organization is large, or if there are turf battles among departments for leadership in user experience design, these sessions can break down barriers. Everyone looks at the set of interview data, discusses interpretations of that data, and agrees on a representation. From this point forward, it's no longer "he said, she said." It's "the data shows this; this is how we interpret it." Subsequent design decisions tend to go more smoothly.

I have experienced one "monkey wrench" in this type of review workshop. On a few projects, new stakeholders were brought in to the mental model review. This was their first real contact with the project and the method. As I began to introduce the mental spaces, these new stakeholders would ask questions about how the interviews were conducted, what the audience segments were, and why there were only six participants from Japan. In each case, the team explained the background, but these stakeholders did not stop there. Having been brought onto a project, they wanted to make a difference, and one way of doing this was calling into question every decision the team had made during the past four to eight weeks. Discussing all this, while worthwhile for the new stakeholders, pretty much derailed the whole workshop. Now, when I learn of new stakeholders joining the team, I set up a *separate* meeting with them before the workshop to explain everything that has happened to this point, to understand their perspective, and to solicit their input. Lesson learned.

What Did You Learn?

Take a long look at the mental model. What stands out? Does the whole mental space seem consistent with what you know of your customers? Which towers are items you would not have thought of before the research? Write up a mental model report or presentation that summarizes your findings.

Decorate the Diagram

Now that the tasks and towers have stopped shifting and the diagram feels stable, you can safely switch from the source document to the diagram. In other words, if any changes need to be made in the future, they will be made directly to the diagram. Now is the time to enrich it with visual representations of the audience segments, regions, voice count, and so forth. I referred to voice count when I showed the first example of a legend a few pages back. Voice count is the representation of how many unique voices went into that particular task (Figure 10.11). You may also want to represent other information in your diagram. If certain tasks or towers represent a single audience segment, you can color the box accordingly.

The colors in this particular legend (in Figure 10.11) example were applied at the tower level. Sometimes I use color in the task boxes to represent audiences or blends between two audiences. In one mental model, I colored consumer tasks pink and business tasks blue, and colored the blended tasks lavender. Yes, these colors ended up looking rather neonatal. Hopefully you have more visual design skill than I.

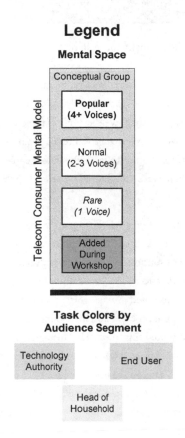

FIGURE 10.11.
Legend with colors added.

Sometimes I put "jewels" in the corners of the task boxes or at the top of the towers to indicate region or audience segment. For several projects in which we conducted international interviews, the client wanted to look at the tasks from a regional perspective. We were looking for towers that were region-specific, or that left out one region entirely. In the diagram below, we considered this tower to represent all three regions, since all three are present to one degree or another in the tasks (Figure 10.12).

Audience Segment Histograms

I have mentioned the company that had the three roles (evaluator, implementer, and maintainer) where we color-coded the towers so that readers could see that one role sometimes would work far into another role's tasks. Since there was so much overlap in the roles supporting each tower and task, we could not simply color the towers just one of three shades. In addition, some of the people we interviewed played two or even all three roles. We used the dots, shown in Figure 10.13, to illustrate the roles involved in each tower. However, there were differing degrees to which the segments supported each tower.

To see these differences, I created a histogram in the spreadsheet that contained the task analysis data. To the right of each tower, I noted the ID numbers of the voices supporting that tower. For each ID, I then filled in a cell with the appropriate role color, for example blue, green, or orange. For ID numbers who represented multiple roles, I filled in a cell with each role color. Then we sorted the colors into a histogram that showed the degree to which each role supported this tower. This histogram helped us understand the blending of the roles from tower to tower, and where one role really took over from the next. What you see below is the middle of the three diagrams, with Evaluator pink showing towards the left and Maintainer blue starting to show towards the right, where these two groups switch over.

Implementer Mental Model Histogram

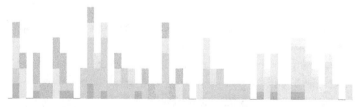

pink = Evaluator, orange = Implementer, green = Implementer/Maintainer
blue = Maintainer, gray = Evaluator/Implementer/Maintainer

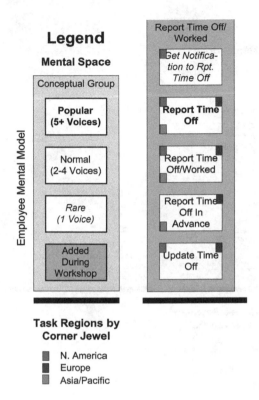

Legend

Mental Space

Conceptual Group

Popular
(5+ Voices)

Normal
(2-4 Voices)

Rare
(1 Voice)

Added
During
Workshop

Employee Mental Model

Report Time Off/
Worked

Get Notifica-
tion to Rpt.
Time Off

Report Time
Off

Report Time
Off/Worked

Report Time
Off In
Advance

Update Time
Off

Task Regions by
Corner Jewel

- N. America
- Europe
- Asia/Pacific

FIGURE 10.12.
Jewels or tiny color chips to represent regions.

In another example, the audience segments varied from tower to tower, so here they are represented by colored dots at the top of each tower, rather than by the tower color itself (Figure 10.13).

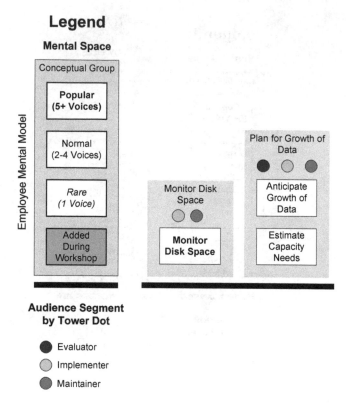

FIGURE 10.13.
Colored dots in towers to represent audience segments.

Essentially, each diagram is different. Figure out what information is important for you to present. Determine the least distracting and most informative way to present it, and experiment.

Ask for Feedback

Once your diagram is finished, and your project guides feel as if they can explain it to others, introduce it to a wider audience. Present your mental model report to as many groups as you can, and ask your team members to do the same. I have seen people print the diagram and affix it to the wall in a high-traffic hallway. I have even seen one client encourage folks to make comments by hanging pencils on strings next to the diagrams.

For now I recommend printing on standard paper and taping the diagram together, instead of spending a lot of money on a professional version. Wait until you have completed the next step: aligning the content and features of your application with the towers in the mental model.

Before addressing the content, though, pause and consider all that you have learned about your audiences. The next chapter shows you how to make any needed adjustments to the audience segments that you started out with.

CHAPTER 11

Adjust the Audience Segments

Compare Results to Original Hypothesis   226
Clarify Segment Names   227
Adjust Segment Definitions   228
Use Audience Segments for Other Projects   232
Transition from Research to Design, Verbs to Nouns   232

Now that you've spent so much time with the participants, combing through their transcripts, you understand these people a lot better. (I often wish I could achieve this level of understanding about my own family and friends!) It's time to compare what you now know to your original hypothesis about the audience segments. Did you get them right the first time around?

Compare Results to Original Hypothesis

Remember that your focus for these segments is the differences between sets of behavior and philosophies. Follow these steps to compare what you know now to your original hypothesis.

1. Dig out your original audience segmentation document. What were the descriptions of the task-differentiated segments in that document? As you read each description, scribble down your gut reaction as to whether that segment name and definition still make sense.

2. Next, open up your recruiting spreadsheet and, on a new tab, cluster the names of the participants who belong to each segment. Think about the people you interviewed in this segment and write down some adjectives and phrases that describe their tendencies. By this time, my memory of "who was who" in the interviews has become fuzzy, so I actually speed-read the transcripts for each participant in one segment. One of the people you classified as belonging to a certain segment may have different tendencies than the rest of the people in that segment. Set this person to the side, along with the appropriate adjectives and phrases.

3. Once you have finished this exercise, compare the new adjectives to the original descriptions of each audience segment. Add your thoughts to the gut reactions you wrote down in the first step.

4. Now, compare the new adjective groups to each other. Are they each distinct? Mark groups that seem similar.

5. Finally, step away from the audience segments and review the mental model. What tasks are interesting, expected, or surprising? Think ahead to the design stage. What differences or similarities might these segments have in usage patterns? Who needs special tools? Write down these thoughts as well.

Now you are prepared to adjust your audience segments.

Clarify Segment Names

In your comparison of your descriptions to the original descriptions, you might decide that a better name for the group will define it more clearly. Usually a better name is easy to determine. If you struggle, then maybe this isn't the right solution. Or maybe it isn't necessary. But if it comes easily, this new name will reflect your deeper understanding of this segment, and it will make this segment clear and distinct from the other groups.

Dow Corning Audience Segments

When we re-assessed the audience segments for silicone manufacturer Dow Corning, much better titles for our audience segments came from the participants themselves. Folks who worked in the lab didn't call themselves "Specifiers." They referred to themselves as "Research and Development." People making products with silicones didn't call themselves "Practitioners." They referred to their jobs as "Production." Those in "Purchasing" more commonly called their jobs "Material Supply." And those ill-defined "Business Clients" were really the "Sales and Marketing" arm of the company that was producing the product.

- Specifier   →   Research and Development

- Practitioner   →   Production

- Purchasing   →   Material Supply

- Business Client   →   Sales and Marketing

With just with a few words changed, our audience segments went from murky to obvious. After this refinement, we developed a persona to represent each of these audience segments.

Adjust Segment Definitions

If, when you compare your definitions to each other, they no longer seem distinct from each other, you need to change your groups. Ignore your original hypothesis for now, and look at the descriptions you just wrote. Start re-arranging the descriptions. Put words that seem to go together in their own groups. Think about the tendencies you heard during interviews. Nudge words around until each group is different and explicitly defined. When you have finished, you can name these new groups. Again, if names don't come easily, perhaps you took a wrong turn. Naming should be the most natural and obvious step of this process.

Figures 11.1, 11.2, and 11.3 are examples of how we have merged and re-defined original audience segments into something more realistic.

For telecommunications provider Bell Aliant, we defined our consumers as folks who might inhabit a household. There were many different tasks, such as paying bills or setting up mobile phone features. We grouped them into five different segments based on these tasks. After talking to consumers, we realized life wasn't that complicated. There were three main roles, any two of which were often played by one person.

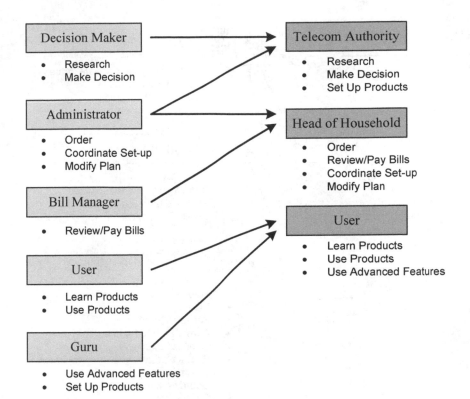

FIGURE 11.1.
A household of telecommunications consumers is not as complex as we thought.

For Sybase, an information management and mobility company, we started off with the idea that customers were mainly split into pre- and post-purchase. Sybase was interested in learning how thought-leaders in the industry went about implementing change within their organizations, so we added that group. And then we split pre-sales into those that already own a Sybase product and those that don't. After the interviews, we laughingly realized our blunder. There were three distinct roles, not one of which mapped directly from our original hypothesis. The complexity we knew existed came in the form of who played how many of these roles.

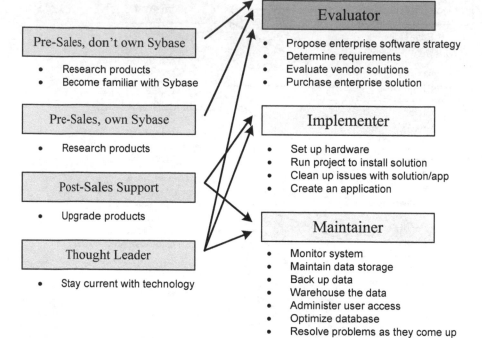

FIGURE 11.2.
At Sybase, a database technology provider, customers don't fall into just one role.

For Engage, an online matchmaking site, we were interested in talking to both singles and people who were already in relationships but interested in helping their single friends find love. Our original hypothesis was close. We knew there were folks who were really outgoing about finding love—both for themselves and for others. We divided these folks into "Social Connectors" and "Good Samaritans." We knew the opposite existed as well—those who were "Self Conscious." We also figured there were people who were less ambitious, who would let love find them. These were the

"Will Participate If Asked." Then there were the people who were trying to increase their odds of success by purposely expanding the types of circumstances in which they might meet someone. After the interviews, it turns out we were right on with the folks who were less ambitious, but we selected a better name for them: "See What Happens." The outgoing folks merged into one group: "Get on the Love Train." But the other two groups splintered into two new groups: They were those who had no plan and were "Trying Too Hard," and those who were apt to examine situations from all aspects, follow rules, and "Think It Through."

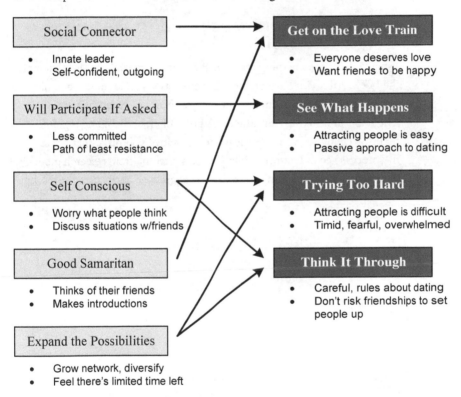

FIGURE 11.3.
Engage singles and matchmakers are re-defined by behavior.

Add your adjustments to the mental model report you wrote at the end of the last chapter.

Use Audience Segments for Other Projects

At this point, you can make a persona[1] for each of your audience segments. You can use the audience segments to write scenarios and use cases. You can draw storyboards using your personas. Because you will need to touch base with your users again in the future to conduct evaluative research, you can use these audience segments to recruit for usability tests or card sorts. The audience segments should have long, productive lives within your organization.

Transition from Research to Design, Verbs to Nouns

Finally, there is something worth noting here at the end of this section of the book. After you have created the verb-based mental model diagram and adjusted your audience segments, you now can revert back to nouns. The habit of thinking from the user's point of view should be ingrained by now. Using verbs up to this point has helped you make that part of the process second nature. Now you are switching from research mode to design mode, so you don't have to be as aggressive about using verbs.

---

[1] See Alan Cooper's *The Inmates are Running the Asylum*, Chapter 9, for the seminal introduction to personas, or read about it here: tinyurl.com/2zygmm See Mike Kuniavsky's *Observing User Experience* Chapter 7 for a how-to description of user profiles. There are other volumes on personas, such as John Pruitt and Tamara Adlin's *The Persona Lifecycle* and Steven Mulder and Ziv Yaar's *The User Is Always Right: A Practical Guide to Creating and Using Personas for the Web*. Many people teach persona creation. Kim Goodwin has worked extensively with Alan Cooper and has a useful interview at the User Interface Engineering site, tinyurl.com/237peu Christine Perfetti of User Interface Engineering writes a good article on the topic at tinyurl.com/ytynyq

CHAPTER 12

Alignment and Gap Analysis

Draw a Content Map of Your Proposed Solution   234
Align the Content Under the Mental Model   237
Consider the Opportunities   248
Share the Findings   254
Print the Diagram   257
Prioritize the Opportunities   258

Now that you have completed the mental model, you can put it to use. You have had a chance to learn more about the people you are designing for. You have illuminated the weak spots in your understanding and removed your assumptions. You are ready to innovate.

One application is to use the diagram as a roadmap for future development. Use it to frame questions about what areas to address and to provide a sandbox for creative ideas, an environment where "luck" and "magic" can happen. Double-check plans to see if they match needs. Because the mental model depicts the whole of the user's environment—it is not focused on one aspect, service, or tool—you can use it to craft a complete experience for the people you interact with. User-centered design becomes an advantage for your organization.

The first step is to align every feature of your solution to the mental model and see where you are.

Draw a Content Map of Your Proposed Solution

I've been going on about mental models for 11 chapters and never mentioned that you should have someone else working on another little project in parallel. This other "little" project is the content map. Someone familiar with your existing and planned offerings needs to sketch them out as boxes that will fit under the towers in your mental model (as shown in Figures 12.1 and 12.2). It's important that the content map is made separately from the mental model, because thinking about all that content forces you to organize it after some fashion, and that framework could unintentionally influence the way you group data in the mental model. Have someone not on your mental model team draw it while you are combing or grouping, or else tackle the content map after you have finished with the mental model.

The content map includes all functionality that already exists and is planned for your solution. Let me assure you that the name "content" does not limit your map to text documents. While it may have started out as a description of an inventory of a news web site or something, it means much more than that these days. Your content map should include all the ways you serve people, including things like monthly account statements or yearly awards banquets, registration for training courses, or a mortgage calculator. Anything that has to do with your relationship with those you serve should be included in your content map.

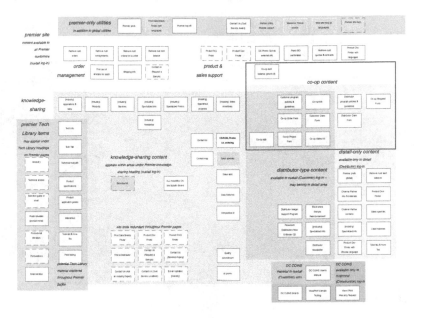

FIGURE 12.1.
Example content map from Dow Corning, with overlapping conceptual areas. (Content "Greeked" for privacy.)

Each content map is the unique creation of its author. One person may draw the same map differently than another person on the same team. It could represent a structure that is sanctioned by the organization, but it's not mandatory. In the end, you will pull out all the boxes from the content map and place them in the mental model, so the structure the author uses does not matter. All that matters is that the width of the box will fit beneath a tower, that you use nouns and adjectives that clearly label the box, and that everyone understands the notation you've used to denote different types of boxes.

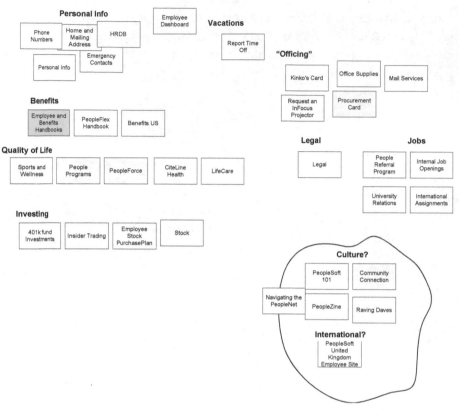

FIGURE 12.2.
A less rigidly organized content map, representing part of PeopleSoft's intranet.

Creating a content map can be pretty intense, especially if you have a lot of ways you serve people. The level of granularity is important. You don't need to detail absolutely everything. If several awards are handed out on a yearly basis, you can just call them "Awards" rather than listing each type. If you have different types of accounts for different kinds of customers, list those separately. You will end up with about 100-175 boxes in your content map. Chiara Fox describes[1] the process she used at PeopleSoft as

---

[1]   Boxes and Arrows article "Re-architecting PeopleSoft.com from the Bottom-Up" by Chiara Fox, June 16, 2002. tinyurl.com/23tqry

first taking an inventory[2] of all the documents the company had about their products. Then she says, "...we created the unified content map. Once the inventory spreadsheets were completed, we were able to pull out the different document types and content types we had found. We identified the larger content areas (e.g., general product information, customer case studies), and then listed the individual examples that existed on the site (e.g., component descriptions, functionality lists)." Chiara's approach is a good way to get at the level of granularity you want. While her inventory and spreadsheet approach was necessary for the content migration she was doing, happily you only will need the higher-level content map.

Align the Content Under the Mental Model

Matching content to multiple towers in the mental model can be a long process. You will pull one box at a time from the content map and study it, determining which towers in the mental model it relates to.[3] You will want to have different project guides with you when you go through this process to help you evaluate things. I have done this with all the project guides present at once, which has the advantage of group discussion and diffusion of ideas. I have also conducted these workshops with just one project guide at a time, covering just the boxes from the content map that are within the project guide's area of expertise. This latter approach has the advantage of not wasting someone's time who is not involved with the content being discussed.

Before you begin, read through all the towers in the mental model and cross off those that are outside your production scope. For example, there were a couple of mental models I developed with towers called "Organize My Time." In each case, my client was not in the business of helping

---

[2] See Jeff Veen's June 2002 essay, "Doing a Content Inventory," on the Adaptive Path web site: tinyurl.com/2bok5j and Janice Fraser's January 2001 article in New Architect, "Taking a Content Inventory," tinyurl.com/2ajacs

[3] An alternate approach to studying one piece of content at a time is studying one tower at a time. You would discuss the meaning of the tower and collect boxes from the content map that supports it. This method requires that you carefully track where the primary location is, however.

people manage their schedules. If my client was Franklin-Covey, the makers of various day planning systems, then we would not have crossed out this tower. This shortcut will make your hunt for matches go a little faster, since your eyes can jump over the crossed off towers.

To make things go more smoothly during the workshops, and additionally engage the project guides as owners of the mental model, I sometimes assign homework. Before each person's appointment with my team, I give her a subset of the content boxes from the map and ask her to try aligning them to the towers that are not crossed off in the mental model. Sometimes I give her a simple list of mental spaces and towers to start with. This homework exercise gives each person a chance to think about the content items ahead of time, as well as a chance to go over the mental model towers several times. It saves the team time during the workshop because many of the ambiguous content boxes have been cleared up. If there is some question what, exactly, "Webcast Event" represents, for example, the project guide can ask the author of the map to clarify its name, or if a content inventory exists, look it up in that document ahead of time.

To begin the workshop, select a content box from the map and make sure everyone understands what it is (Figure 12.3). Discuss its intent and its value to the organization and the user. For example, the content box "Theaters Near Me" is a tool that looks up a zip code and lists the theaters within a 10-mile radius. You look at the mental spaces of the moviegoer mental model and decide "Choose a Theater" is the most likely time a person will want to see this list of nearby theaters. A tower called "Choose Easy Theater" seems to be a good primary match for this content item.

Choose a Theater

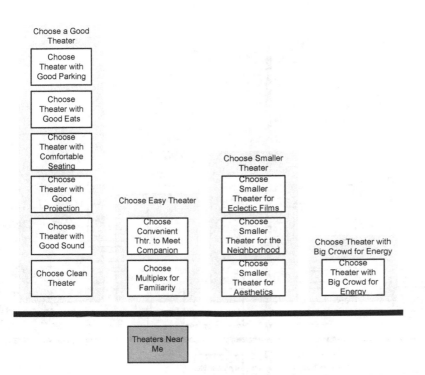

FIGURE 12.3.
Slot "Theaters Near Me" under "Choose Easy Theater" as its primary location.

You look for other matches, and see "Choose Screening Time" in the "Choose a Time" mental space. This is a good secondary location (see Figure 12.4). You look for others. "Avoid Being Late to Theater" in "Go to the Movies" is a reasonable match, if you think that the person is choosing a theater that is fastest to reach.

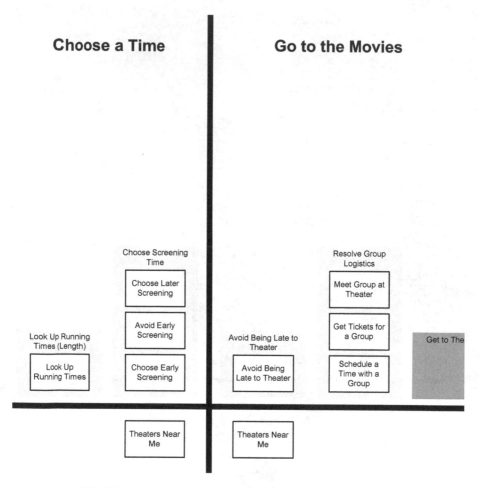

FIGURE 12.4.
Place "Theaters Near Me" under two other towers as secondary locations.

You might wonder how strong the link is between "Theaters Near Me" and "Avoid Being Late to Theater." Perhaps only one person on your team suggested it. Do you keep it in the diagram, and later risk thinking that this tower is fully supported? If it is marked as a secondary match, and if this tower only has secondary matches beneath it, you do not run the risk of assuming it is fully supported. Secondary matches act more like contextual marketing. If a person is doing one thing, you might want to remind her or make her aware of this content you just slotted. Amazon does this online with their "Customers who bought this item also bought" list for each product. There is a fine line to walk here. Sometimes these contextual links are really, truly helpful. More often than not, though, they make users suspicious of your motives. In this example, though, the utility of having the distance to each theater listed or the driving time would be wholly appropriate.

You might have content that does not match any tower in the mental model. Usually this is because the item just isn't anything that people want when trying to do what they're doing. A great example is the ubiquitous stock ticker that came free with certain portal software, so it ended up on the home page of many intranets—despite the fact that usage statistics showed that employees were not interested in the Dow Jones when they were logging in to record their vacation days. Granted, at some point in the day some employees might want to check stock prices on a holding, but that activity is part of a different mental model. It did not come up in our interviews within the scope of studying how an employee interacts with her company. Checking stock prices is supported by a different application—most likely her investment web site. On the other hand, if your usage statistics on the employee intranet show that people do check the stock ticker a lot, and it wasn't mentioned in any of your interviews, then you have a little more exploring to do. Ideally you would have noticed the stock ticker statistic during your secondary research review before drafting your interview prompts.

Then again, maybe this content matches a tower that is missing from the mental model. You can add a tower (sans tasks) to the diagram if it is something that members of your team have heard from customers before. For example, in the moviegoer content map there are two boxes labeled "Write Fan Mail to an Actor" and "Write Fan Mail to a Director." There was no matching tower in the mental model, so the team added a tower labeled "Write Fan Mail." (See Figure 12.5) This tower represents a behavior the team has observed in moviegoers before.

Identify with a Film

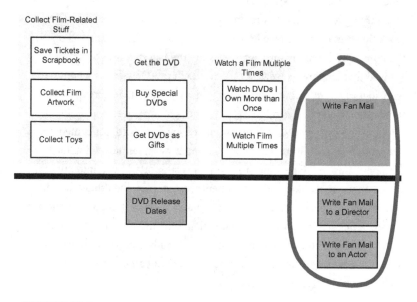

FIGURE 12.5.
Content that does not match an existing tower. Add a tower based on experience.

Another tougher example illustrates the selection of primary versus secondary locations. "DVD Release Dates" is a list of movie titles nearing release on DVD, for consumer purchase or rental. You look at the mental spaces and immediately gravitate to "Watch a Film at Home" as its primary space. The tower "Rent a Film" seems to be a perfect match. Consumers wanting to rent a film will want to see when the film they wish to rent will be released, so you slot "DVD Release Dates" beneath it, marked as a primary location. Is there a tower that deals with purchasing a DVD? Yes there is, in the mental space "Identify with a Film," the tower "Get the DVD." But wait, which tower is the primary location? Buying or renting the DVD? You will need to more explicitly define the content, or change it, to make this decision. Perhaps your stakeholders see sales as their business, not rentals, in which case you would make the "Get the DVD" tower the primary location. What if you redefined the tool to allow people to pre-order upcoming DVDs? Then it would be primary under "Get the DVD" again. If you skip ahead in the content map, you'll find two boxes called "Rent a DVD from Sponsor" and "Reserve a DVD from Sponsor." These could be primary under "Rent a Film" and "DVD Release Dates" could be secondary, as a contextual reference which may be helpful to people (see Figure 12.6). Precise definition of your solutions in line with your organization's goals helps you focus and cleans things up for your users. *Every piece of content should have one and only one primary home.*

Watch a Film at Home

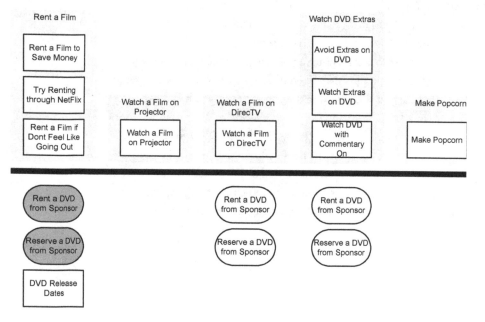

FIGURE 12.6.
Knowing that your organization is in the business of selling, not renting films helps you determine that "DVD Release Dates" is secondary under "Rent a Film."

One more difficulty I have encountered in alignment workshops is a tendency by project guides to misinterpret a mental space as being the same as a previously held concept. It's often my fault for choosing similar vocabulary to what they use internally because I heard people using those words. For example, the mental space "Evaluate a Product" means that a person specifically spends time talking to others about a product, comparing it to other products, testing out the product, evaluating the manufacturer, narrowing the list of providers, getting a quote, and calculating the costs. The phrase "Evaluation" had already been in use in that particular organization, however, and it meant choosing a product. Testing a trial version of the product was a separate phase from "Evaluation," and the group kept tripping over the meaning of that mental space. In a related example, the tower "Ask Someone to Research Products" was misinterpreted as "discuss options with knowledgeable person," not "assign the product evaluation project to someone else" as the mental model intended. During our alignment workshop, I had to keep an eye out for content that was slotted according to the internal understanding of these two concepts (Figure 12.7).

Content Map and Alignment Shortcut

A quick and dirty way to skip the content map and conduct an alignment workshop is to look at your existing solutions while checking off towers in the mental model that have content supporting them. Make a different kind of check mark for towers that have content that tangentially supports them (secondary content).

FIGURE 12.7.
Project guides slotting content on the mental model at Microsoft
by using a whiteboard for towers and sticky notes for content. Each
person used a different sticky note color and slotted content under all
the towers, then the group went tower-by-tower to discuss what really
fit and what was primary. (Photo by Carey Wilkins of VML.)

Logistically, you can approach aligning content the same way you
approached the mental model review, with a projector and a remote
connection for those not in the room with you. Cut boxes from the content
map and paste them into the mental model. By cutting the boxes instead
of copying them, you can track your progress through the content map
more easily. Make sure you save an original copy, first. If your team is
co-located, another fun way to slot the content beneath the towers is to use
sticky notes. Post the printed mental model to the wall. Write[4] a content
item on one color sticky note, then find its primary home in the mental
model and stick it underneath that tower. Write the same content item on a
different color and stick it underneath a secondary tower.

---

4 Or print the content to printable sticky notes. For details, see my note at the end of the section
   "Comb Early and Often" in Chapter 9.

Notation for Aligned Content

When you show the mental model to others, you will want an easy way to spot the "primary home" for each piece of content beneath the towers. Agree upon some type of notation that makes the primary locations show up easily at a glance. Decide on this notation up front with your team because changing the notation halfway through the exercise is a pain in the neck. The notation that has worked well for me is to use a box with a fill color for the primary, and a box with light or white fill for the secondary locations.

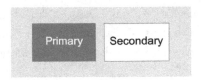

You will also want to easily pick out planned and suggested content. I have used different colors to represent this, although when selecting colors, I need to be careful that they appear distinct to someone who is colorblind. I have also seen people use different colors to represent certain categories of content, which means I can't use color to indicate existing, planned, or suggested content. In this case I use thicker border lines to indicate the values.

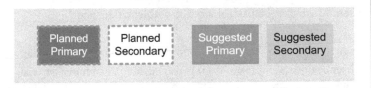

You may even want to call out content that you don't own. I have used a different shape for this in the past, such as an oval or a box with rounded corners.

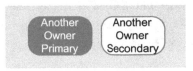

Repeat with this second color for all secondary placements. After you are finished, transcribe the sticky notes to the electronic version of your mental model.

Since mental models are very wide when printed out, a nice aid to the process is to create a cheat sheet with all the mental space and tower names printed vertically. Usually the names will fit on two sheets of paper. This cheat sheet is a good reference for project guides present at the workshop, and it is also useful, as I mentioned, while doing slotting homework.

During slotting, you will make many decisions about content. These do not have to be final decisions. You will have another chance to refine ideas and discuss alternatives during the design and implementation of the particular solution.

Consider the Opportunities

After you have aligned all the content, you will have a diagram that looks like a city skyline along a lakeshore, with a distorted reflection below it. You will want to sit down with all your project guides in another workshop to analyze the gaps you see in this reflection and see what opportunities are apparent. A concise report of these opportunities is what you want to present to your executives, along with a potential timeline.

The first thing to look at in your workshop is the obvious gaps where there is absence of content items. Your hope is that you can find a gap that you can fill pretty easily. For example, in the moviegoer mental model, there is a gap under "Collect Film-Related Stuff" in the mental space "Identify with a Film." See Figure 12.8. This is a really easy opportunity for JMS Entertainment to partner with a toy store selling action figures, or eBay listings for movie posters. On a web site, they could simply post links or ads next to each movie that features some of the available collectibles.

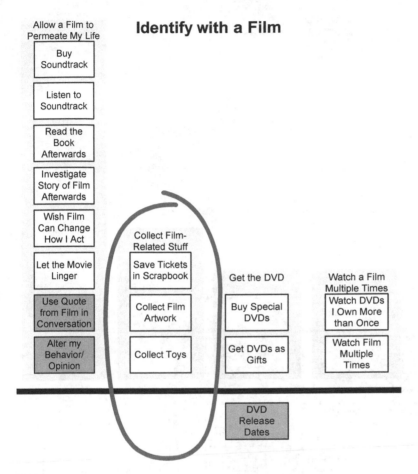

Identify with a Film

FIGURE 12.8.
A gap exists below "Collect Film-Related Stuff."

The second thing to look for is scarcity of content items. Think about where you can flesh things out a bit. In the moviegoer mental model in Figure 12.9, there is not much content under the tower "Watch a Certain Genre During Winter" in "Choose Films." There are secondary placements of "Film List," by genre and related films, but nothing else. Here is a perfect opportunity to do something new, like maybe associate

Choose Film

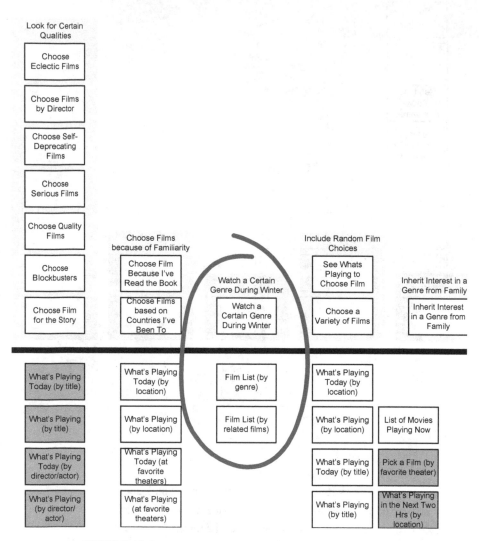

FIGURE 12.9.
There is no primary content below "Watch a Certain Genre During Winter."

films with a season. JMS Entertainment could market films based on the time of the year, such as, "Have you seen your summer blockbuster yet?" or "Get in the mood for Halloween—watch a horror film!" Write these ideas as opportunities in your report.

The third thing to look for is opportunities to redefine, combine, or augment existing content. Even if there is primary content beneath a tower, can you do better? For example, you might discuss that "DVD Release Dates" ought to really be a few things. Perhaps people would prefer a way to look up a consumer release date for any movie title, regardless of when it was released, not just upcoming releases. Real movie buffs might use a tool that looks up one film's release date and compares it to other titles that were released around the same time. You could augment this tool with data from DVD sales of those films and link it to press about the film. Nearly anything is possible, as long as it makes sense for the moviegoer and for JMS Entertainment.

Hand-in-hand with this brainstorming is a fourth opportunity. You might see synergies between towers that aren't represented by the aligned content or relationships between mental spaces. You might look at two mental spaces and say, "Hmm, these two mental spaces have the same stuff supporting them. Why is that? Is there an opportunity for us to empower the user some way?" You can put these two mental spaces together in some way that people don't really expect, but is going to make things easier for them. For example "Choose a Theater," "Go to the Movies" and "Eat Dinner" could be supported by a tool that lets a person park the car once and walk between the theater and the restaurant, saving the frustration and cost of parking twice—not to mention the savings to the environment. By looking across towers and across mental spaces, you might see the next "killer product" waiting for you.

On the flip side, there will be gaps in the mental model that you will not want to get into as an organization. I mentioned the tower "Organize My Time" at the beginning of this chapter as an example of a tower to cross out because your organization is not in the business of helping people manage schedules. Now during gap analysis, you will want to go over these crossed-out towers a second time to double-check your decisions. Suppose you crossed out all the towers in the mental space "Watch the Film" because JMS Entertainment can't be in the theater with the moviegoer (see Figure 12.10).

But two towers jump out as opportunities: "Compare Film to Book" and "Look for Jokes/Reference." Couldn't JMS Entertainment provide this information for the moviegoer ahead of time? A composite of quotes from published reviewers about how the film measures up to to the book would be easy to create. Certainly a list of "in-jokes" or "Easter eggs" would attract people's attention. It could be on a web site; it could be on a poster in the theater's game arcade; it could appear in advertisements on bus stops. Do any of these make sense as an opportunity to pursue? This is the kind of idea-generation and discussion that the mental model can elicit.

Then again, maybe the tower you crossed out should stay crossed out. The tower "Inherit Genre from Family," visible in Figure 12.9, is the kind of thing JMS Entertainment sees no way to facilitate. It's a gap they just don't want to get into the business of supporting.

"Do We Want to Be in That Business?"

Gap analysis is a great way to ask yourself, "Do we want to be in that business?" of each tower. Use this chance to brainstorm ideas, or eliminate a certain tower because it just isn't right for your organization. Conduct this exercise once a year, to help your team stay on track.

Finally, gap analysis can be used towards other ends. You can create a content map of your competitors' offerings and align them to your mental model in overlays. Take a look at those gaps between what you offer and what they offer, and see if there are any opportunities there.

The gap analysis report that you create after this workshop will make known the biggest gaps in your solutions, the most well-covered areas, and ideas your team has generated around these areas. This report, as it stands, will be of great interest to your executives and investors. You can go one step further, though, and pick out the highest priority concepts to implement.

Watch the Film

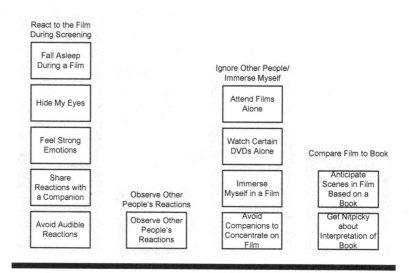

FIGURE 12.10
The mental space "Watch the Film."

Share the Findings

Now is the time to let everyone in your organization know what you have discovered. Print the mental model with the slotted content and post it anywhere you can get permission. Assign a person neighboring that location to be the emissary of the mental model, explaining the findings to interested parties. To encourage people to propagate the mental model in presentations and documents, you could make a separate image of each mental space that people could paste in to their presentation, with a reference to a version of the whole diagram posted somewhere internally. Smaller chunks of the mental model will be more readable and digestible.

Watch the Film

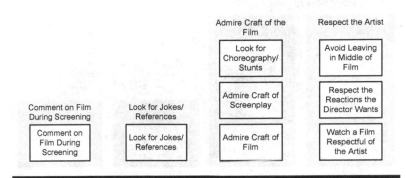

I have heard of project guides spending a few weeks presenting parts of the diagram to various sub-groups, making sure everyone can take advantage of the collected knowledge.

I have invited executives to presentations 15 minutes earlier than other folks, so they can stand in front of the diagram on the wall and walk from left to right, asking me questions as they go (see Figure 12.11). As I answer their questions, I explain how it will be used to direct product design.

This kind of walkthrough is quick, to the point, and stays in the context of "missed" and "future" opportunities that executives usually focus on. *Many executives have told me that they've never before seen all this information collected so succinctly in one place.* Jeff Veen, User Experience Manager at Google and founding partner of Adaptive Path, said that showing the mental model diagram to his directors was a good political tool for getting resources, "It was, 'Wow, these designers have a rigorous process for something we don't really understand. Let's give them what they need.'" His directors recognized the diagram as something that would shine light on a poorly understood problem set and gave the team the green light on their design.

FIGURE 12.11.
Mental models posted on conference room walls. Photos, clockwise: Natsukari-san presenting the diagram at Dow Corning Tokyo; Dan Arganbright with diagram at Quixtar; Laura "Lad" Decker at Microsoft; CEO Suneet Wadhwa with team (Alicia Dougherty-Wold, Indi Young, Karen Wallace) at Engage.

Print the Diagram

Printing the diagram presents a challenge. Usually, the diagrams are several feet long. You can go the "crafty" route, print it on ordinary paper, and tape it together. You can go the deluxe route and print on a color plotter, if your company owns one. You can have an outside printing firm print it in color, or even laminate it. Getting the diagram in a format that a plotter can accept, however, requires patience and negotiation skills. If you are sending the file to a printing firm, these firms and their employees are usually *unable* to recognize files in Visio or Omnigraffle format. File formats such as PDF or JPEG are acceptable. Often, you have to divide your file into sections. I've never heard of someone getting a print made in less than four days (more on this on the book site at www. rosenfeldmedia.com/books/mental-models/content/resources).

In order to get your file into PDF format, you will need Adobe Acrobat Professional. When you install it, icons should appear in your Visio or Omnigraffle toolbar. Create your PDF file by clicking the "Convert to Adobe PDF" icon, uncheck the "Include Custom Properties" checkbox in the window that appears, hit "Continue," and choose "Flatten all layers." A PDF file will be generated over the next 30 seconds or so, and, if you default to showing the file, it will appear on your screen in "Fit Width" mode, so your entire mental model appears as an unreadable long strip. Reset to 100% view and you will be able to read it. Ordinarily the width and height of the file appear in the footer of the PDF window, so you can tell your printing vendor how long and wide the diagram is. I've seen diagrams run from 12 to 25 feet, 8.5 inches high, with 6-point font. Or there are diagrams 44 inches high that are about 20 feet long, with larger fonts. Choose what works best for you.

Label	Name	Description
Toys	Toy Store Collectibles by Film	Ads for collectibles related to the film from partner toy stores
eBay	eBay Collectibles by Film	Links to collectibles related to the film on eBay
Seasonal	Seasonal Marketing Campaigns	"Have you seen your summer blockbuster yet?" or "October is the month to see a horror film!"
DVD Release	DVD Release Date Lookup by Film	Enter film title or choose from list filtered by director, actor, other? See the DVD release date.
DVD Co-Releases	DVDs Released Around Same Time as Film	List of other DVDs released within default 14 days of a film. Can adjust timeframe.
DVD Sales	DVD Sales Numbers	Sales of DVDs to date for selected film.
Press	Film Press	News, press releases, reviews for selected film.
Parking	Park Once, Easy Walk	Enter film to see or theater to attend plus restaurant or cuisine, see best parking location to walk to both locations. Extend to parking reservations?
Comparison	Reviewer Comparisons to Book	Summary of comparisons to the book from reviewers.
Jokes	Jokes & References	List of in-jokes and "Easter eggs" for selected film, for feature and

FIGURE 12.12.
Prioritization spreadsheet with priorities assigned to the four columns. The Label column is to identify points on the chart. The Description column reminds you, six months down the road, what the idea meant.

Prioritize the Opportunities

When you look at a list of all the items you thought of during gap analysis, 80% of them always seem to look really important. It can be difficult to arrange them. Different people will have varying opinions as to the importance of the items. When I was working with Testmart in 1999, the Vice President of Engineering, Dave Eisenlohr, introduced me to his version of the Six Sigma prioritization matrix. In 2001, Janice Fraser[5] introduced me to a chart that helps visualize the matrix. Together these techniques help whittle down your list of "do first" ideas to the truly significant.

---

5 See Janice Fraser's April 2002 essay "Setting Priorities" on the Adaptive Path site.
tinyurl.com/yt3h7o

		(4 = really feasible)		(4 = really important)	
Technical Difficulty	Resource Availability		To the Business		To the Customer
2.00	3.25		3.50		1.00
1.75	3.00		2.50		1.00
3.75	3.50		3.50		1.00
2.75	2.75		2.00		1.75
2.75	2.75		2.00		1.50
2.75	2.75		2.00		1.00
3.75	2.75		1.00		1.50
1.00	1.25		3.00		4.00
3.75	3.25		1.00		3.00
2.75	2.00		1.00		4.00

Begin by writing each idea listed in your gap analysis report into rows
of a spreadsheet. Then make four columns to the right of the ideas and
label them:

- Technical Feasibility
- Business Feasibility (time, people, cost)
- Importance to the Business
- Importance to the User[6]

---

[6] Use whatever labels make the most sense for your organization, and you can add columns
if you like.

Label	Name	Description
Toys	Toy Store Collectibles by Film	Ads for collectibles related to the film from partner toy stores
eBay	eBay Collectibles by Film	Links to collectibles related to the film on eBay
Seasonal	Seasonal Marketing Campaigns	"Have you seen your summer blockbuster yet?" or "October is the month to see a horror
DVD Release	DVD Release Date Lookup by Film	Enter film title or choose from list filtered by director, actor, other? See the DVD release
DVD Co-Releases	DVDs Released Around Same Time as Film	List of other DVDs released within default 14 days of a film. Can adjust timeframe.
DVD Sales	DVD Sales Numbers	Sales of DVDs to date for selected film.
Press	Film Press	News, press releases, reviews for selected
Parking	Park Once, Easy Walk	Enter film to see or theater to attend plus restaurant or cuisine, see best parking location to walk to both locations. Extend to
Comparison	Reviewer Comparisons to Book	Summary of comparisons to the book from reviewers.
Jokes	Jokes & References	List of in-jokes and "Easter eggs" for selected film, for feature and DVD.

FIGURE 12.13.

Prioritization spreadsheet with averages of the columns.

You will be filling in the cells for each column with a value of 0-4,[7] where 0 represents not feasible/important and 4 represents really feasible/important (Figure 12.12). With the input of all your project guides, assign a number to each idea along this scale. Getting input from your project guides can take several forms. You can hold another joint workshop where everyone discusses the values assigned to each item, or you can contact people individually. In my experience, the former is more powerful. It not only ensures that the numbers assigned are reasonable, but also puts the team on the same page with regard to near-term direction. I recommend getting everyone together for a prioritization workshop.

Next, add a columns for "Average Feasibility" and "Average Importance." Set the formula for each row for "Average Feasibility" to the average of the "Technical" and "Business Feasibility" columns. Do the same for "Average Importance," using the "Importance to the Business" and the "Customer" columns (Figure 12.13). The chart (Figure 12.14) will graph these two numbers to provide a visualization of which ideas to pursue.

---

7   Again, you can use whatever range you want. I have found that people start bargaining in quarter points, so I like to start the range at 0 and include enough numeric elbow room to negotiate.

OVERALL SCORING		(4 = really feasible)		(4 = really important)	
Average Feasibility	Average Importance	Technical Difficulty	Resource Availability	To the Business	To the Customer
2.63	2.25	2.00	3.25	3.50	1.00
2.38	1.75	1.75	3.00	2.50	1.00
3.63	2.25	3.75	3.50	3.50	1.00
2.75	1.88	2.75	2.75	2.00	1.75
2.75	1.75	2.75	2.75	2.00	1.50
2.75	1.50	2.75	2.75	2.00	1.00
3.25	1.25	3.75	2.75	1.00	1.50
1.13	3.50	1.00	1.25	3.00	4.00
3.50	2.00	3.75	3.25	1.00	3.00
2.38	2.50	2.75	2.00	1.00	4.00

Usually the average will turn out to be exactly the same number for several of your entries. This indicates further discussion is needed. Ask your team to look at the ideas in question and verify or adjust their numbers in the four Feasibility/Importance columns. (Do not adjust the two average columns, since those are the calculated columns.) This exercise can spark some useful insights and debate.

To create the chart of the two average columns takes a little know-how, so sketching it by hand isn't a bad option. If you are using Microsoft Excel,[8] you will need to use a series to graph the numbers. It is my understanding that the labels for a series don't show up on the graph in Excel 2003, so you need to enter a separate series for each row to make the labels show up. On a second worksheet, insert a graph. Right-click on the graph and select the Source Data menu item. Click the Series tab in the resulting pop-up window. You will see a scrolling list in the bottom left of the pop-up that shows the labels for each data point. Select the label you wish to explore, and the associated cells will appear in the Name, X Values, and Y

---

[8] I have posted a Microsoft Excel template of the prioritization worksheet in the Resources section of the book site. ✿ www.rosenfeldmedia.com/books/mental-models/content/resources

Values fields in the bottom right of the pop-up. Clicking the Add button beneath the scrolling list will add a label called "Series1," and you can set the associated cells for the Name, X Values, and Y Values from the prioritization spreadsheet. The only odd thing about this workaround is that each data point on the graph will have a different icon for the dot itself (see Figure 12.14).

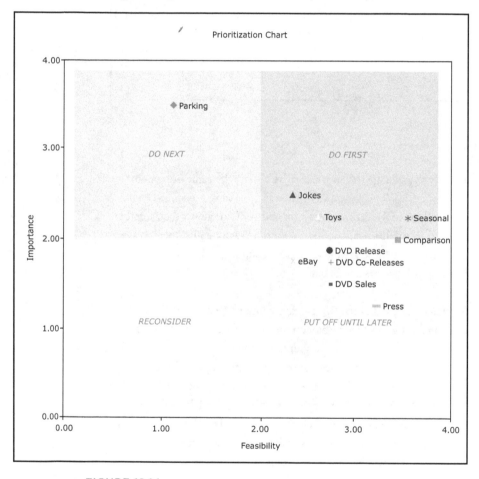

FIGURE 12.14.
Prioritization chart showing the first three ideas you should implement (Jokes, Toys, Seasonal), and the fourth idea to start next (Parking).

You can easily see that Jokes & References, Toy Store Collectibles by Film, and Seasonal Marketing Campaign are the ideas to implement first. They are important and feasible. Park Once, Easy Walk is a more involved project that is also important to the moviegoer, so that project should be planned next. But before lighting out on these top projects, be sure to confirm their value. Do the same for the items that fall into the lower-right quadrant. Something like Reviewer Comparisons to Book might get elevated in the pipeline depending on how the details of the project look. You will want to get agreement and approval from all parties involved, as you usually would before beginning a project.

Once you have approval, create a project timeline and budget following your normal procedure. Present this timeline and the gap analysis report to the executives and investors.

One of the things that often crops up in prioritization is the need for better architecture, especially in software products. The next chapter will reveal how to derive architecture from the mental model.

CHAPTER 13

Structure
Derivation

Derive High-Level Architecture   266
Provide Vocabulary for Labels   276
Test Your Structure and Labels   280
Generate Features and Functionality   280

Getting the architecture of your product right is sometimes an elusive pursuit. Often you will come to one conclusion, and after implementing it you learn of a requirement that just doesn't fit what you have. As far as web sites are concerned, many corporations have made it an unintended habit during the past decade to re-architect their product sites every two years. Obviously this cycle of constant redesign can be a drain on finances, patience, and respect. Conversely, manufacturers of hard goods purposely redesign their products on a cyclical basis to increase customer interest. However, innovation is difficult to guarantee. The mental model can alleviate these frustrating conditions. Develop your design based on the strategy and vision that the diagram represents. By matching the structure people naturally use in practice, you can set up a lasting foundation for iterating your product.

In the introduction I said that I use the word "design" to indicate digital, physical, and environmental interactions that people carry out to accomplish something. It means that you are making something for someone to use. That something can be a digital product, a physical object, or a space in which to obtain a service, or anything else you can conceive. Going forward, I encourage you to merge my suggestions in this book with established techniques in your field.

Derive High-Level Architecture

The highest two levels of architecture are easy to see now that they just about fall out of the mental model diagram. The process of deriving them takes three hours at most. Start by writing down your highest-level groups—the mental spaces. The idea behind this exercise is to look for affinities among any of your mental spaces. I like to use a medium where I can drag the text around and put it next to other labels. However, if you're a whiteboard or pen-and-paper person, those will work too. You can draw all sorts of lines between the labels. You will probably have 20-50 mental space labels to start with. Then you group them. In Figure 13.1, 38 mental spaces become 11 clusters.

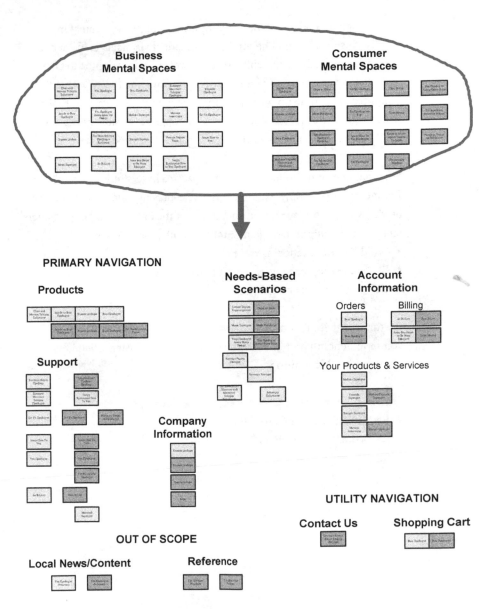

FIGURE 13.1.
The clusters show that there are close relationships between the mental spaces, thus a similar architecture can be used for both audiences. (Content "Greeked" for privacy.)

If your mental model only has, say, 14 mental spaces, you can skip the affinity clustering step. The total number already is manageable enough to translate directly into architecture. However, it doesn't hurt to take a look. Perhaps one or two mental spaces will have an affinity with others. Your goal is to come up with a list that is manageable in length. Eight is reasonable. Ten is reasonable. Even 18 or 20 are reasonable because you will have a chance to cluster the labels again in the product design.

So, what does it mean for one mental space to have affinity with others? Let me show you some examples. In a role-playing game, such as those on the Xbox, the PlayStation, or any one of the online worlds, the gamers' mental model might include mental spaces like the following:

- Build My Reputation as the Best
- Become Recognized as the First to Do/Discover
- Celebrate an Accomplishment

All three of these mental spaces have "community recognition" as a common denominator and can be grouped into a cluster called "Earn Community Recognition."

Other gamer mental spaces need to be considered, too:

- Be Helpful/Generous to Others
- Be Mean/Harass Others

Both of these mental spaces also involve community recognition, but they don't really cluster organically. The former is positive recognition and also includes a personal sense of altruism. It is not purely community-based. "Celebrate an Accomplishment" in the first list also exhibits this characteristic. However, "Be Mean" is negative recognition, and a reputation for meanness can become quite pervasive within a community. After this observation, I might group the mental spaces like this:

- Earn Community Recognition
  - Build My Reputation as the Best
  - Become Recognized as the First to Do/Discover Something
- Promote Good Will
  - Celebrate an Accomplishment
  - Be Helpful/Generous to Others
- Undermine the Progress of Others
  - Be Mean/Harass Others

Thus five mental spaces now become three. As you work through the rest of the gamer mental spaces, you may find that "Earn Community Recognition" and "Promote Good Will" end up in the same cluster anyway, because there is not as much distinction between them as there is between, say, "Earn/Find Gold" and "Make My Avatar Look Interesting."

In the moviegoer example, I start with 14 mental spaces, study these, and create six clusters (Figure 13.2). The first labels I cluster together are those that pertain to selecting the film and when and where to see it. This all has to do with decisions and logistics. Included in this cluster is the concept of choosing a theater for a lively crowd and getting a group together to go. I call this cluster "What, When, Where, & with Whom."

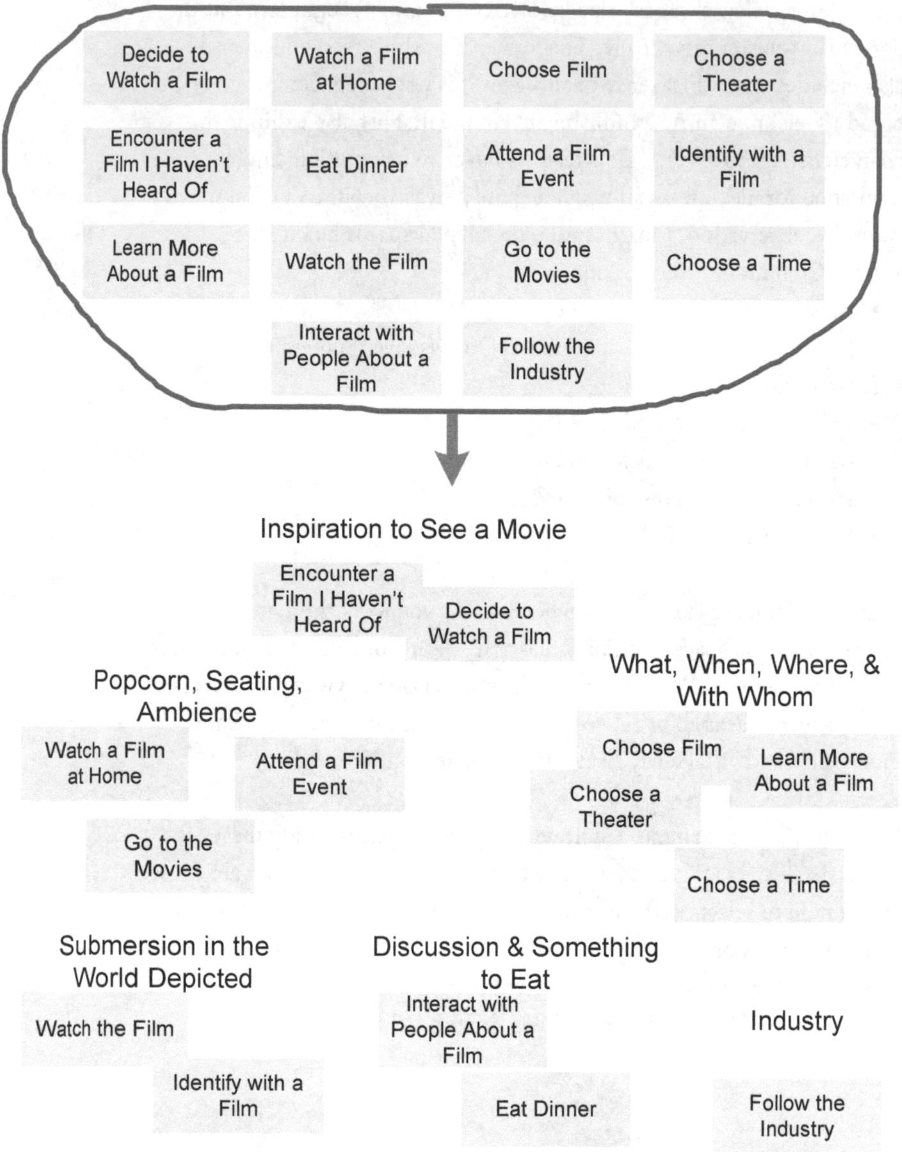
Decide to Watch a Film
Watch a Film at Home
Choose Film
Choose a Theater
Encounter a Film I Haven't Heard Of
Eat Dinner
Attend a Film Event
Identify with a Film
Learn More About a Film
Watch the Film
Go to the Movies
Choose a Time
Interact with People About a Film
Follow the Industry
Inspiration to See a Movie
Encounter a Film I Haven't Heard Of
Decide to Watch a Film
Popcorn, Seating, Ambience
Watch a Film at Home
Attend a Film Event
Go to the Movies
What, When, Where, & With Whom
Choose Film
Learn More About a Film
Choose a Theater
Choose a Time
Submersion in the World Depicted
Watch the Film
Identify with a Film
Discussion & Something to Eat
Interact with People About a Film
Eat Dinner
Industry
Follow the Industry

FIGURE 13.2.
Moviegoer structure derivation.

The second cluster I see is about settling in to view the film—finding your seat, finding the right venue if you're at a film festival, or popping your own popcorn if you're at home. These all have to do with setting up your preferred movie-viewing environment. I call this cluster "Popcorn, Seating, Ambience."

The third cluster is about having the idea to see a movie in the first place. I labeled this cluster "Inspiration to See a Movie."

After I gather these first three clusters, I look at what's left. There is a mental space about dinner, one about the industry, one about interacting with people, one about watching the film, and a fifth about identifying with a film. It seems that the dinner and people interaction mental spaces cluster because people talk about the film over food, so I label them "Discussion & Something to Eat." For the remaining three I have to look at the mental model diagram again to understand more completely what they represent. I look at the mental space "Identify with a Film." It is about watching the film multiple times, letting it permeate your life, and collecting related items. I look at "Watch the Film" and see that it is about experiencing the film, understanding the message, and reacting in conjunction with companions. Each mental space is slightly different, but they both deal with the impact of the film on you, your life, and your companions. I decide to cluster them as "Submersion in the World Depicted," although submersion might not be exactly the right word. These are just "working" labels, so it's okay to move on.

The last mental space remaining is "Follow the Industry," which is distinct enough from all my other clusters that I keep it separate. Thus, I end up with six clusters.

So, what about these clusters? Should all the cluster names be verbs, like the mental spaces? It's not necessary, and often this is a good place to switch to nouns, if it feels natural. The clusters represent the way you put your product together. For a running watch, it might be the number of buttons needed along the rim of the watch and the displays on the face.

For an investment institution, it might be the levels of advice you offer customers, the information in a monthly statement, and the functions customers use to manipulate their money. For a software application, it would be the menus.

Over the past years, I have been using this technique to define navigation for a lot of web applications and sales sites. When I say navigation, I mean that these clusters represent primary, utility, and footer navigation—just the universally visible links on a web site. Secondary navigation, which shows up depending on which primary item you select, can be derived from a clustering of the towers in the mental spaces represented. The third level of navigation, if there is one, can be a subset of the towers depicting various tasks. Lower level architecture, such as lists of product specifications, should be organized using traditional information architecture approaches,[1] by looking at the topics and grouping them accordingly.

In the moviegoer example, first I will look at the content aligned beneath the mental spaces in my clusters and make sure that there are no unsupported clusters. For example, the cluster "Inspiration to See a Movie" will happen before a person comes to this web site, so I will leave it out of the navigation of that web site. I might see some interesting synergies that I didn't catch during the gap analysis. For example, putting the local restaurant links in the same cluster with the film rating tools makes me think about a face-to-face discussion group tool, where someone can post ahead of time that they'll be at the Royal Ground Coffee House after the 9PM showing of *Blade Runner* at the Rafael Theater, and if you care to join him, he'd be happy to discuss the film with you and your friends. Keep your eyes open for thoughts like this, and even if they're unlikely to be high priorities, it is worth adding them to the mental model and your priority list.

---

[1] How could I resist mentioning *Information Architecture for the World Wide Web* by Louis Rosenfeld and Peter Morville? Also see the seminal paper about the "scent of information," in "Using Information Scent to Model User Information Needs and Actions on the Web" by Xerox PARC researchers Dr. Ed Chi, Peter Pirolli, Kim Chen, and James Pitkow. tinyurl.com/2h466e

The next step is to think about the clusters in terms of where they fit in the universal navigation. Some of the clusters might belong in the footer or the utility navigation. A store locator tool that clusters by itself might belong in the footer, as it does by convention for many web sites. Or, if you have a cluster for the purchasing process, such as a shopping cart and checkout flow on a web site, it belongs in the utility navigation. Create two navigation items in a new diagram called "Footer" and "Utility Navigation" and move these clusters next to them.

The next step is to come up with working nomenclature for the primary navigation labels for the rest of your clusters. I do this by going to the mental model and writing down a list of the kind of content that belongs in each cluster. In the moviegoer example, the content under the mental spaces belonging to the "Submersion in the World Depicted" skews towards the "Identify with a Film" mental space, since "Watch the Film" has no content beneath it. The content under "Identify with a Film" follows the theme of DVD release dates, collectibles, and fan mail. I ask my team to consider these things together, and we come up with the working label "Fan Zone" for this navigation item. Figure 13.3 depicts the other labels the team recommended.

I re-name the first one "What's Playing Where" to be more understandable, and rearrange the resulting temporarily-labeled navigation items:
- Recommendations & Discussion
- What's Playing Where
- Get In There
- Fan Zone
- Industry Insider

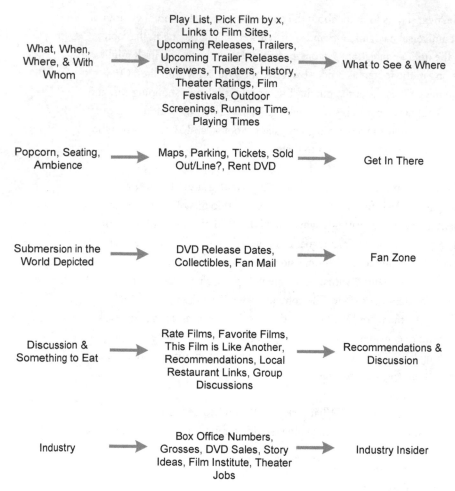

FIGURE 13.3.
Select a working label based on the kind of content associated with each cluster.

There are connections between many of these areas, such as picking a movie and getting a ticket, or picking a movie and recommendations. These connections will be developed within each section, just like teleportation booths[2] that can jump you from your office to the cafeteria or from the cafeteria to the human resources department on the top floor of your building. But the five primary sections of the web site provide the moviegoer a hesitation-proof set of choices. The hallway test applies here. The moviegoer was walking down the hallway thinking to himself, "I'm supposed to meet Nick and Michelle for a movie tomorrow afternoon at the Lark Theater. I wonder what is playing." He'll see the "What's Playing Where" section and, since it maps directly to his purpose, he will click it without delay. If a different moviegoer was walking down her hallway thinking, "I'm taking the kids to the multiplex for their play date tomorrow; I need to get tickets and figure out where to have the other moms drop off their kids," she will see the "Get In There" section and proceed therein immediately. There is no, "Well, where do I go from here?" And there should be no dithering between two options, as there is on some airline sites that have both "Book Travel" and "Reservations" links.

Deriving architecture from the mental model this way can seem kind of cut and dried. It would be ideal to roll it out just as you see it. In reality, other pressures can contort the structure you have obtained. Powerful stakeholders frequently demand entries that seem redundant or concepts that live at lower levels in the mental model. Your job is to try to talk them out of their recommendation by showing them how you took the architecture from the way people actually behave. Show them all the steps you took. Discuss your interpretations with them and remain open to differing analyses. Sometimes you will win, and sometimes you will lose.

---

[2] Too bad these don't exist yet. Well, imagine something like a booth-version of the transporter room in *Star Trek*, labeled for a specific destination, and you'll understand the usefulness of the analogy.

The final step is to sketch how your solution will look. Sketch the product structure without any of the content details, if you can (Figure 13.4).

FIGURE 13.4.
Sketch the product structure without content details. This sketch is for the JMS Entertainment moviegoer web site, using temporary labels for navigation.

Provide Vocabulary for Labels

If your product will have text on it, such as labels on a washing machine or in the menus of a software application, you'll want to work on transforming those temporary working labels into final labels now. Since you have been trying to preserve the vocabulary you heard in the interviews, this step may be a short one.

Put your temporary working labels into the first column of a matrix, and fill the second column for each label with alternate wording for the item (as shown in Figure 13.5). Brainstorm these alternate labels with your team and look up wording in the transcripts. Then decide on the best label you want to use and write this label in the third column.[3]

You may have two labels that you think will perform equally well. Keep them both. The next section will show you how to address this situation.

---

[3] Also see Peter Merholz's best-practices brief, "How Labels Affect Usability and Branding," at Adaptive Path. tinyurl.com/2hbfu6

WORKING LABEL	SUGGESTIONS	BEST LABEL(S)
Test & Produce	~~System~~ Test and Production Support	Test and Qualify
	~~System Test and Production Services~~	**Test and Qualification Supports**
	Product Evaluation	
	~~Tech Support~~	
	System Production Data	
	System Test Data	
	~~Design Resources~~	
	Samples and Test Data	
	~~Proving Ground~~	
	Real World Use	
	~~Integrate~~	
	~~Evaluate~~	
	Resolve Issues	
	Test and Production	
	Test and Production Resources	

FIGURE 13.5.
Write your working label in the first column. Gather nomenclature ideas based on vocabulary from the interviews in the middle column. Choose the best label and put it in the last column, modifying it to fit a noun-based naming scheme.

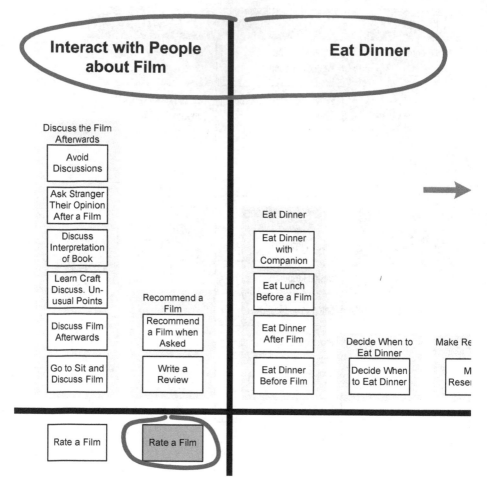

FIGURE 13.6.
Start with the feature "Rate a Film." Refer to the mental model to be sure you understand the user's intentions.

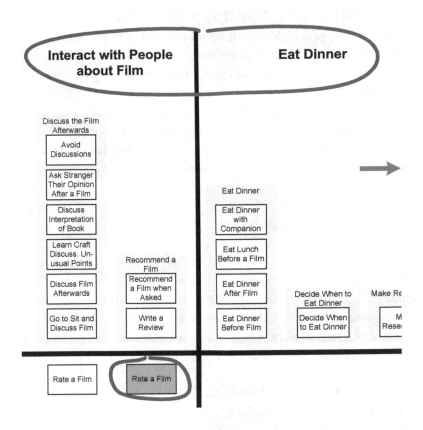

Test Your Structure and Labels

Once you have your high-level structure delineated and labeled, it's a great time to double check your decisions with real people in the field. You will want to use a closed card sort[4] to verify that the structure you come up with will make sense to those who will use it. List each of your sections, and provide a pile of potential content for participants to assign to these sections during the tests. After each test, solicit input on how the participant defined each category in her mind, to double-check that the word you used carried its intended meaning.

If you have two labels for one item that you want to decide between, assign one label to half the tests and the other label to the other half.

After conducting the tests, analyze the data and your notes to see if you should make changes to the structure or labels you set up. The card sort method is straightforward, but it requires a lot of time to set up and conduct the tests and much concentration during analysis. Allow two weeks elapsed time for the testing. You should be able to overlap recruiting with the brainstorming of vocabulary.

Generate Features and Functionality

When you begin work on the highest priority project that you defined after gap analysis, you will most likely begin with a series of sketches of the functionality you want to achieve. You may wish to capture these sketches in documentation with annotations and present them to team members and stakeholders to verify whether the design meets your organization's goals. This documentation goes by various names depending on your field of work: functional requirements, schematics, product specifications, wireframes,[5] etc. Before you begin sketching, though, refer to the mental model and *jot down the user's intentions that this particular feature supports.*

---

4 See Donna Maurer's upcoming book on card sorting, published by Rosenfeld Media.
    www.rosenfeldmedia.com/books/cardsorting

5 For web site design, Dan M. Brown writes a good description of creating wireframes in Chapter 10 of his book *Communicating Design.*

The PeopleSoft Consulting Group

Back in 2001 I created a mental model and navigation derivation for People-Soft's public web site. Back then PeopleSoft sold enterprise software for accounting and human resources. The mental model we developed had a mental space called "Implement Software" just after the point where a customer made a purchase decision. "Implement Software" contained towers that had to do with planning, customization, installation, and initial training. The derived navigation for the public site had a section for rolling out the software called "Implement."

The existing web site that we were replacing had a section called "Consulting." There was a large division within the corporation that provided consulting to customers. This division primarily helped customers adjust the software to their workflow and implement better workflow in the first place, but the consultants also helped people decide which combination of People-Soft and partner products would work best for the customer's organization. This consulting division actually brought in more than half of PeopleSoft's revenue for many consecutive quarters.

On the existing site there was also a list of PeopleSoft partners who either supplied add-on software or who also consulted in a manner similar to the above. Our mental model indicated that customers thought of both the internal consulting division and external partners as relatively equal resources to help with implementation. We suggested combining the two existing areas under "Implementation." This made perfect sense from the customer perspective. However, the Executive Vice President of the consulting division felt that putting his division and the partners on equal footing on the web site would threaten a critical revenue stream.

We showed him that customers didn't necessarily think "hire consultants" when they were trying to "choose an enterprise system." In the end, says Camille Sobalvarro, project lead at the time, "We capitulated. This is an example of how the balance between user preference and business drivers is sometimes achieved." We added "Consulting" back to the primary navigation. Furthermore, in the interest of selling cycles on the new server farm, we also relabeled "Implement" as "Application Hosting."

Look at the cluster this feature belongs to in the structure, and study the towers of those mental spaces to understand the mindset of the user. Then pinpoint the feature's box in the diagram and look at the primary tower it supports. Write down the tasks in this tower and some words describing the user's mindset (as shown in Figure 13.6). Also look at the secondary towers it supports to see if there are any additional things it should include. Finally, list your assumptions about the environment the user is in, such as, "There is no guarantee that other community members are online at the same time as this person." You should end up with a couple of paragraphs to help you focus your design.

The idea is to keep your feature design simple and sleek. You want to deliver precisely what the user is seeking. This is not to say you should make simplistic designs, because users often prefer the efficiency of accomplishing many related things at once. But make sure that these features are related, from the same cluster of mental spaces in which you started.

If you're collaborating with your team for this exercise, a return to sticky notes is helpful (see Figure 13.7). The first steps can be accomplished by moving around sticky notes with tasks, towers, and content written on them. The final design can be sketched, accompanied by the description of the User's Intentions and the Assumptions.

FIGURE 13.7
David Poteet, President of New City Media, mulls over features for each part of an application at the Imperial College of London with Tom Miller, Director of Communications, left. They wrote down tasks or task groups from the mental models on sticky notes and put them onto posters that represented key interactions, like "Apply to Imperial" or "Research at Imperial." (Photo courtesy of James Vogt of New City Media.)

After your team agrees on the direction, make a prototype or a paper prototype[6] and test it with real people.

You know how to do the rest.

---

[6] See Carolyn Snyder's book *Paper Prototyping*. www.snyderconsulting.net

Index

SYMBOLS

30-Day Cycles 38
3M Printscape Personalized Note Kit 194

A

About Face 2.0: The Essentials of
   Interaction Design (Cooper and
   Riemann) 32
active listening 121
adjusting audience segments 59
Adobe Acrobat Professional 257
affinity clustering. See clustering
affinity diagrams. See also mental model
   diagrams 3
affinity game (Sesame Street) 166
agile development 35
Agilent Technologies, Inc. 13, 20
alignment and gap analysis 234-255
      content alignment under mental
         model 237-244
      content maps and 234-235
      notation for aligned content 247
      prioritizing opportunities 258-
         259
      sharing findings of 254-255
alignment workshops 237-241
analysis, of transcripts. See transcripts
   analysis
Apple 16
applications, organization of 14,
   266-280
applications for mental models
      alignment and gap analysis
         234-254
      structure derivation 266-279
Arganbright, Dan 256
assembling topics for research interviews
   98-99
assumptions about user's environment
   282
atomic tasks 188
audiences, complicated 51

audience segment histograms 220
audience segments. See also task-based
   audience segments
      core attributes of 59
      differences in mental models of 12
      keywords for, in interviews 99
      names of 54-59
      task differences among 210-211
      in traditional marketing 46
automatic coloring of mental model
   diagrams 200
automatic method for mental model
   creation 198-200

B

Becker, Lane 117
behavior affinity, and audience
   segmentation 50-53
Bell Web Solutions 180
bias, in interviews 97, 110
Bierut, Michael 10
big fan (audience segment) 57
Bloug (Louis Rosenfeld's blog) 65
brainstorming sessions 34, 49
Brown, Dan M. 280
business, design as advantage in 18

C

card sorting 26, 280
cash, mental model of 24
characteristics (personal) 73
cheat sheet for mental model diagrams
   248
Chen, Kim 272
Chi, Ed 272
chronological order, in mental models
   136, 211
Circumstance (of tasks) 134
clarity, of design strategy 16
closed card sorts 280
clustering of mental spaces 266-276
cognitive research 8
Cohen, June 49
coloring, of mental model diagrams 200
comber (team role) 189, 193

combers vs. interviewers 194
combing transcripts 132–136, 192–195
companionship (core attribute of movie-
    goer audience segment) 60, 62., 64
complaints 134, 136
compound tasks 146, 172
Consorti, Simonetta 20
construction of mental models,
    overview of 4
constructive research 29
content alignment 237–248
    adding content-based towers 242
    example of 239, 240
    gap analysis and opportunities 248
    primary locations 243
    primary vs. secondary locations
        243
    secondary locations 240
content map and alignement shortcuts
    245
content maps 234–236
    for Dow Corning 235
    granularity of 236
    for PeopleSoft 236
controlled vocabularies, for task tagging
    188
conversations, interviews as 111
Cooper, Alan 32, 69
core attributes of audience segments 59
core teams 42
costs
    of interview recruiting 92
    for transcribers 127
craft (core attribute of movie-goer audi-
    ence segment) 62
customer-first philosophy 20
customer feedback 26
customer goals vs. product preferences
    109
customer perspective, importance of 22

D

Decker, Laura 256
decoration of mental model diagram
    218–221
De Muro, Jacqueline 13, 20
design
    as business advantage 18
    confidence in 9, 10, 11, 12, 13
    mental models and decisions in 12
    simple vs. simplistic 282
    successful iv
    use of term viii
Designing for Interaction: Creating
    Smart Appications and Clever Devices
    (Saffer) 97
desires 134, 136
diagrams
    mental model of typical morning 4
diaries 32
Difficult Conversations (Stone, Patton
    and Heen) 110
dinner party conversation 107
dismissing interview participants 122
    122
Disney Concert Hall (Los Angeles) 11
documentation, of features and
    functionality 280
Doughtery-Wold, Aicia 256
Dow Corning 235, 256
drawing programs 200
dry erase sheets, portable 180
Duncan, Craig 14, 20, 41

E

Eisenlohr, Dave 258
electronic files, formatting tasks in 148
electronic task affinity grouping 185
electrostatic dry erase sheets 180
emotion, in user experience 17
empathy 2
empirical research 29
enabler (audience segments) 58
Engage.com 75, 256

evaluative research 26
Excel (Microsoft) 183, 200, 261
executives, presentation of mental model
    to 255
expectations 134
experience, as product 18
experience strategies 17, 19
explanations (of tasks) 134
exploratory research 29

F

feature definition 14, 280–283
feelings 134
fees
    for interview recruiting 92
    for transcribers 127
field visits 32
film purist (audience segment) 57
findings, from mental model 218–220
fitting text in mental model diagram boxes
    200
following the conversation (interview
    technique) 111
footer navigation 273
Fox, Chiara 140, 236
Franklin-Covey 238
Fraser, Janice 258
Freitas, Ryan 15
functionality, of products 280–283

G

gap analysis. See alignment and gap
    analysis
gap analysis report 253
Garrett, Jesse James 19
Gehry, Frank 11
generative research 26
ghosting, of interviews 120–121
good tasks, examples of 136–140
Google Analytics 12
granularity
    of content maps 236
    in interviews 113, 122
    of tasks 135
greeting, for interviews 100

Gross, Terry 108
grouper (team role) 189
grouping, of tasks. See also task affinity
    groupings
        in audience segment definition
        process 50

H

Hackos, JoAnn 121
hallway test 135, 275
headings, in Word, use in task affinity
    grouping 181–184
Heen, Sheila 110
high-level architecture, of structure
    derivation 266–276
high-level tasks 135
hit rate, for interview recruiting 91

I

immediate experiences 112
Imperial College (London) 283
implied tasks 134, 140
in-person interviews 106–107
information design 198
instructions to individual task combers
    193
interaction design 14
internal vocabulary, problems of 244
international dialing 85
international interviews 125–127
international teams 149
interpreters. See translators
interview bias 97
interviewer (team roe) 189
interviewer bias 97, 110
interview granularity 113
interview participants
    dismissing problem participants
        122
    real-life story of 90
interview recruiting
    audience segment and demographic
        goals for 74–76
    costs for 92
    database hit rate for 91

demographics for Engage.com 75-76

importance of personal characteristics for 73

methods of 88-92

one person representing multiple audience segments 73

participants not fitting requirements 89

participant stipends 85

real-life story of 89

schedule coordination 88

specifying details for 72-84

spreadsheet for 72, 73

use of pop-ups for 90

interview recruiting screeners 76-80

audience segment definitions 80

audience segment identification 82

beginning of 77

completed form 78

description of 76

essay question for 81

logistical questions on 83

speaking ability, gauging 82

interviews

assembling topics for 98-99

as conversations 107, 111

conducting 106

examples of 116-121

initial greeting for 100

keywords for 98

nervousness during 104

non-leading 97, 107-108

open-ended questions 122-129

prompts for 96, 103

real-life story of 117, 118

recording of 100, 104

reminders for 99

softball questions for 101

uncovering tasks in 114

interviews. *See also* interview scope; interview techniques; transcripts

interview scope

assembling topics 98-99

interview reminders 99

prompts 96, 103

recordings 100, 104

research goals, setting 94-96

interview techniques

active listening 121

avoiding discussion of tools in 112

consideration for participants 114-115

following the conversation 111

ghosting 120-121

immediate experiences, focus on 112

in-person versus remote 106-107

open-ended questions 110

practice interviews 113

product preferences vs. customer goals 109

six rules for 109-113

thinking during interviews 116

transcripts, importance of 127-128

tricks 121

user's vocabulary, use of 110-111

interview transcripts. See transcripts

intranets 214, 236

introductory documents, for international interviews 126

iPod 16

J

jewels (segment designators) 219-221

JMS Entertainment 248, 251, 252

K

Keeley, Larry 18

keywords, for interview themes 98

knowledge preservation 24

Kuniavsky, Mike 89

L

labels 273-274, 276-278

Landelle, Sarah 180

landing pages 215

Laurel, Brenda 97

leading questions 122-129

leads, for projects 40, 41

legends, for mental model diagrams 207

levels, in mental models 96
Life (computer game) 173
lifespan of mental models 7, 23
looking for patterns. *See task* affinity
   groupings
luck 10

M

Macy's 179
make believe artist (audience segment)
   58
Maurer, Donna 280
mediums (tools) 134
memory, organizational 24
mental model creation 198–220
   automatic method for 198–200
   block-by-block diagram creation
     200–206
   diagram decoration 218–221
   feedback on mental model diagram
     223
   process overview 8
   reviewing model with project guides
     216–217
   splitting model into multiple
     diagrams 210–215
   summary of findings 218–220
mental model diagrams
   building, block-by-block
     200–206
   cheat sheet for 248
   chronological order in 136, 211
   coloring of 200
   creating towers in 203–205
   decoration of 218–221
   example of 6
   format for 198
   getting feedback on 223
   legends for 207
   printing of 257–258
   review of 216–217
   splitting of 210–215
   of typical morning 4
mental model process 8
mental model research, qualitative nature
   of 96

mental models. *See also* applications for
   mental mondels; mental model creation; metholdology for creating mental
   models
   description of 2
   reasons for using 9
   use of term 5, 8
mental spaces
   differences among 172
   explanation of 4
   formatting of 177–184
   high-level architecture and
     266–272
   of role-playing games 268
   task affinity groupings and 170
Merholz, Peter 17, 276
method for creating mental models
   definition of task-based audience
     segments 46–61
   interviewing participants 106–129
   mental model creation 198–220
   shortcuts in 38
   specifying recruiting details 72–84
   task affinity groupings 164–190
   transcript analysis 132–159
Microsoft. *See also* Excel; Visio; Word
   content alignment project at 246
   mental model presentation at 256
Miller, Tom 283
miscommunication 121
money, mental model of 24
mornings, typical 4
Morville, Peter 272
movie-goer mental model. *See* task
   analysis; task-based audience segments
movie buff (audience segment) 56
MP3 players 16
multiple diagrams, for mental model
   210–215
Myers-Briggs psychological types 46

N

Nagai, Deborah 43
names
    of groups for task-based audience
      segments 54–59
    of performer-task sets 55
Natsukari-san 256
navigation, mental space clusters and
    272
nervousness during interviews 104
New City Media 21
non-leading interviews 97, 107–108
    six rules for 109–113
normal tasks 207
notation, for aligned content 247
no words of your own (interview
    technique) 110–111

O

Omnigraffle 198, 200
online resources
    bibliography 4
    footnote references 4
    for international dialing 85
    introductory documents for
      international interviews 126
    prioritization worksheet template
      261
    Python script for creating mental
      model diagrams 199
    recruiting spreadsheet 73, 195
    World Clock Meeting Planner 85
open-ended questions 110, 122–129
opportunities
    from gap analysis 252
    prioritization of 258–259
    prioritization spreadsheet for
      258–259
    resulting from gap analysis
      248–252
organizational change 19–20
organizational memory 24
outlines for task affinity grouping
    180–185

Outline view, in Word 180–185
overlapping tasks 211

P

participants. *See* interview participants
particular tasks 135
patterns
    grouping tasks into 164–174
    shifting, in task affinity groupings
      173–176
Patton, Bruce 110
PDF format, for mental model diagrams
    257
PeopleSoft 236
performer-task matrix 52, 54
performers of tasks 53
personality types 46
personas 32, 69
philosophy (of tasks) 134, 135
Piontkowski, Mary
    on identifying audience
      segmentation 51
    mantra 10
    on mental model shortcuts 36
    Word template for sticky-note
      printing 194
Pirolli, Peter 272
Pitkow, James 272
planned content, notation for 247
politics, mental models as mediators in 14
popular tasks 207
portable whiteboards 180
Porter, Michael 18
Post-its 194
Poteet, David 21, 283
practitioners, on project team 41
preference research 26, 27, 28
preferences, task vs. 134
primary navigation 273–276
printing
    of mental model diagrams 57–258
    on Post-it notes 194
Proctor & Gamble 6
product architecture 266–280
product design 21

product managers 40
product preferences vs. customer goals 109
product testing, user research for 26
project guides 42, 120–121, 216–217, 237
project leaders 40, 41, 94
project managers 40
project plans 189
project practitioners 41
project stages 34
project support 43
prompts, for interviews 96–98, 103
prototypes 283
Python script for creating mental model diagrams 198, 199

Q

Qualcomm 89
Quixtar 256

R

Radio Shack, phone recording devices 101
rare tasks 207
reading mental model diagrams 211
real-life stories
    of audience segment histograms 220–228
    of combing and grouping 195
    of in-person interviews of media buyers 107
    of interview participants 90
    of interview recruiting 89
    of miscommunication from pronunciation differences 121
    of PeopleSoft web site 281
    of recruiting 43
    of research interview 117, 118
    of scholarly phrases 154
    of tasks database 188
    of transcriptions, problems with 129
    of use of customer verbs 23

of value of mental models 13, 14, 15
reasons for using mental models
    assembling original ideas 12
    avoiding politics 14
    avoiding unending cycles of redesign 15
    clarity in direction 16
    confidence in designs 9, 10, 11, 12, 13
    deriving architecture 14
    distinguishing among different solutions 12
    evolving organizations 19
    overview 9
    paying attention to the whole experience 16
    strategy continuity 23
    using design as a business advantage 18
    validating that ideas match needs 13
recordings
    software for 101
recruiters 43, 89, 125
recruiting one person representing multiple audience segments 73
recruits tab, of interview screener 87
redesign, non-ending cycles of 15
Redish, Janice 121
regions, in mental model diagrams 219
reminders, for interviews 99–102
remote vs. in-person interviews 106–107
research goals 94–96
research interviews. See interviews
research scope, for task-based audience segments 65–67
reviewing mental model with project guides 216–217
Riemann, Robert 32
role-playing games 268
roles, for task affinity grouping teams 189–190
Rosenfeld, Louis 65, 272
rules for interviews 109–113
Rutter, Kate 180

S

Saffer, Dan 97
San Disk MP3 player 16
scalability, of mental models 34
scenarios 33
Schauer, Brandon 18
scholarly phrases 154
scientific vs. intuitive methods 11
secondary locations, in content alignment
   240
segmenting complicated audiences 51
self-created mental models 37-38
Sesame Street affinity game 166
shifting patterns in task affinity groupings
   173-176
shortcuts, in method
   overview 33
   summary of 38
Six Sigma 258
Snyder, Carolun 283
Sobalvarro, Camille 15, 281
social movie-goer (audience segment) 56
softball questions, for interviews 101
splitting mental model into multiple
   diagrams 210-215
Spool, Jared 16
spreadsheets (electronic), for task affinity
   grouping 180, 183-185
stakeholders 94, 95-96, 216
statements of fact 134
sticky notes
   organizational change and 20
   printing on 194
   use in content alignment 246
   use in feature and functionality
     design 282
   use of, to track tasks 177
stipends, for interview participants
   85-87
Stone, Douglas 110
story (core attribute of movie-goer
   audience segment) 60
strategy, continuity of 23

structure derivation 266-279
   features and functionality, creation
     of 280-283
   high-level architecture 266-276
   product labels 276-278
   testing structure and labels 280
styles, for Excel cels 200
Sybase 15, 43, 73

T

tagging, of atomic tasks 188
task-based audience segments 46-61
   adjusting 59
   as basis for personas 69
   description of 46
   differences in behaviors 46
   distinguishing tasks, behavioral
     affinity grouping of 50-53
   distinguishing tasks, listing of
     48-50
   distinguishing tasks, naming groups
     of 54-59
   setting research scope on 65-67
task-performer matrix 52, 54
task affinity groupings 164-189
   additional interviews, need for 185
   combers vs. interviewers 194
   forcing a structure 165
   formatting towers and mental spaces
     177-184
   grouping tasks into patterns
     164-182
   identifying likenesses and
     differences 165-166
   left-over tasks 185
   mental spaces and 170
   numbers of tasks 177
   order of steps 164-168
   parking lots in 174
   shifting patterns in 173
   sticky notes, use of 177
   team roles for 189-191

tasks. *See also* transcript analysis
   audience segments and differences in 210-215
   behavioral affinity groupings of 50-53
   brainstorming sessions for 48-49
   combing transcripts for 132-136, 192-195
   compound 146-148, 172
   databases for 188
   definition of 133
   examples needing work 140-148
   feelings and 134
   formatting of 148-152
   frequency of 136
   good examples of 136-139
   high-level 135
   identification of 133-136
   implied tasks 134
   listing of distinguishing 48-50
   list of performers of 53
   particular tasks 135
   philosophies of 134, 135
   rare, normal and. popular 207
   third-party tasks 134
   titles of, fitting in diagram boxes 200
   uncovering in interviews 114
   verb + noun format 136
teams
   international 149
   participants in 40-43
   roles for task affinity grouping 189-190
templates
   for printing sticky notes 194
Testmart 258
third-party tasks 134
three C's 9
tools
   avoiding discussion of, in interviews 112
   in tasks 136
topics, for research interviews 98-99
towers
   adding, based on content 242
   formatting of 177-184

tracing quotes back to transcripts 137
tracking desires and complaints 136
transcribers 43
transcript analysis 132-159
   combing for tasks 132-136, 192-195
   compound tasks 146-148, 172
   formatting tasks 148-152
   practice examples 154-159
   tasks, examples needing work 140-148
   tasks, good examples of 136-139
   verbs, importance of 153-154
transcripts
   combing for tasks in 132-136, 192-195
   importance of 127
translators 81, 88, 126, 127
Tufte, Edward R. 198

U
United Nations
   International Strategy for Disaster Reduction 14
   Prevention Web Team 20
universal navigation 273
university effect 127
user's intentions 282
user's vocabulary 110-111
user-centered design 30
User and Task Analysis for Interaction Design (Hackos and Redish) 121
user research 26-29
uses of mental models viii, 6, 30, 113
utility navigation 273

V
vague verbs 146
VALS Personality traits 46
value of mental models 13, 14, 15
Veen, Jeff 12, 21, 237, 256
verbs 22-24, 136, 146, 153-154
videos, use as scenarios 33
Visio (Microsoft) 198, 200
visual language, mental models as 11
voice count 207, 218

W

Wadhwa, Suneet 256
Wallace, Karen 256
Walt Disney Concert Hall (Los Angeles)
   11
whiteboards, portable 180
who, what, when, where, why, and how
   (open-ended questions) 110
whole experience 16
why questions 115
Wikipedia 173
Word (Microsoft) 180–183
word processing documents, for task
   affinity grouping 180–183
workshops
     alignment workshops 23 244
     in mental models development
       process 7
     on priority optimization 260
World Clock Meeting Planner 85
world time clocks 85

Y

yamaha.com 214
Young, Indi 256

ACKNOWLEDGMENTS

I would like to thank the following people for their influence on the mental model method as it continues to mature over time:

Mary Piontkowski, who is not afraid to point out where I might be wrong-headed. Thank you.

Gina Davis and Paul Stiff, who saw the strengths of early mental models and encouraged me to develop the method.

Jim Hook, who told me that what I thought was common knowledge might not really be as common as I supposed.

For beta-testing draft chapters and clearing up confusion: Craig Duncan, Simonetta Consorti, Andrea Villa, Jen Richardson, Isabelle Peyrichoux, Ingrid Hart, Marcus Haid, Ginny Redish, Todd Wilkens, Brandon Schauer, Ryan Freitas, Bjorn Hinrichs, Jay Morgan, Jon Littel, Lauren Kessler.

Clients who believed in the possibilities: Camille Sobalvarro, Yen Lee, Suneet Wadhwa, Karen Wallace, Joelle Kaufman, Lee Thompson, Terie Clement, Joel Arabia, Karen Semyan, Carey Wilkins, Renae Gottschall, Paola Dovera, Mark McCormick, Secil Watson, Georgina Corzine, Connie Frennette and Clare Barr and Kathy Parsons (the "Kh" team), Srinivas Raghavan, Hanneke Krekels, Simon Parker, Jim Mier, Tim Adams, Jackie De Muro, Bob Bebb, Lakshmi Sundar, Dan Arganbright, Suzanne Van Cleve, Bryan Vais, Eric Fain, Tom Gruber, Peter Ostrow, Maclen Marvit, Enrique Alvarez, Matthew Nelson, Chad Carson, Christina Avrett, and Kevin George, who was the first person to say, "You should write a book about this," while we were riding in a taxi out of Pasadena, in September 2000. Real examples make all the difference when explaining something. I want to thank all my clients who gave me permission to publish our work.

Teammates Donelle Gregory, Evelyn Wang, Mark Phillips, Carolyn Snyder, Laurie Bell, Chris Baum, Jen Klafin, Nate Bolt, Genelle Cate, Jean Anne Fitzpatrick, Darcy DiNucci, Adele Framer, Dave Covey, Chiara Fox, Sarah Rice, Kate Rutter, Lane Becker, Mike Kuniavsky, Peter Merholz, Jesse James Garrett, Janice Fraser, and Jeff Veen. And Gary Wang, who wrote the first Python scripts that spit out the diagrams.

I could not have done this without all the good people involved in production at Rosenfeld Media: Karen Whitehouse for inflating my ego. It's her fault that I'd like to write another book. Peter Morville for sending me an email out of the blue offering to do a technical review; I can just imagine him sitting down at his desk that morning thinking, "Gee, I have nothing pressing to do. I think I'll offer to read Indi's manuscript." Right. Liz. Danzico and Allison Cecil for taking everything off my hands and making my part in the production process blissfully easy. Marc Rettig, whose comments on the "mental models versus alignment diagrams" debate, as well as the book's positioning, were quite helpful early on. Lou Rosenfeld for having the determination to start Rosenfeld Media and do something right. It is pure joy to work with him and watch his publishing philosophy unfold. I am confident that his efforts will force a sea-change on the industry.

I owe my sanity during the writing of this book to my friends: my mom, Gus Young, graduated high school in 1958, and won a four-year scholarship to a university to study math. If she hadn't turned it down in order to help support her seven siblings, I am convinced she would have been one of the first software engineers. Yes, her name is Gus. My brother, Greg Young, for urging me to write a book, any book. Philip Ramsey for urging me to continue writing after dinner while he did the dishes. Carolyn Wan for making me laugh with email about her three young daughters. I wish I could tell stories like she does. Lucy Simon for moral support and distracting workouts at the pool. Matt Stephens for all the phone calls when I was frustrated with words. Marjorie Forman for keeping me from being too much of a recluse. Carey Ritola for celebrating my 15 minutes of fame.

And for her constant presence curled up napping on my desk: Megahertz (13 January, 1988–21 June, 2007), my 19 year old cat who was really a space alien, I'm sure.

—Indi
29 June, 2007
San Rafael, California

ABOUT THE AUTHOR

Photo by Lucy Simon

Solving people's problems and inventing reliable methods to produce solutions have been Indi's passions since youth. After earning her BS in computer science from Cal Poly in 1987, Indi took some graduate courses at Colorado State University in Fort Collins until she discovered that a master's in Computer Science would only set her up to be a good compiler writer. Instead, she set her sights on solving the ease-of-use problem inherent in all software of the day. Oddly enough, she began by designing the graphical interface to an edit-compile-debug environment written for a Unix-based supercomputer. After that, she pursued the more user-friendly pen-based arena, designing several handy applications for the various start-ups involved, such as Go and Slate. When AT&T killed the pen-based GO OS in 1992, Indi moved on to an operating system that had captured the majority of users and had a thriving developer community. (Yes, Windows.) Luckily, the Web came along soon thereafter.

Indi began her work in Web applications in 1995 as a consultant in interaction and navigation design. She worked with clients who were incredibly sharp, including folks at Visa, Charles Schwab, ADP, Autodesk, and loads of dot-com start-ups, including, most famously, WebVan. A founding partner of Adaptive Path in 2001, she continued her consulting practice with equally erudite clients at Sybase, Agilent, Dow Corning, Microsoft, PeopleSoft, and Qualcomm. Indi specializes in deeply understanding the people for whom her clients design applications, nearly always employing research and mental modeling methods.

Indi, a resident of San Rafael, California, is a sixth generation descendent of the small group of Californian pioneers who trudged back through the snow to rescue the Donner Party. Having subsequently gotten into quicksilver mining industry in Silicon Valley, these forebears set the technology tone for the family. Accordingly, Indi played with nascent computers as a kindergartener and is now a lifetime member of the Association for Computing Machinery (ACM).

CPSIA information can be obtained
at www.ICGtesting.com
Printed in the USA
JSHW031934161222
35015JS00003B/3

9 781933 820064